Beyond Pills, Knives & Needles

CAMS Technology

The Next Step in the Healthcare Revolution

Treating the real cause of disease and ill health (disturbances in the human energy field) using scalar energy, quartz crystal technology, and the resonant frequency of the Earth to solve Healthcare problems.

by

CHARLES J. CROSBY, DO, MD(H)

The information, ideas, and suggestions in this book are not intended as a substitute for medical advice. Before following any suggestions contained in this book, consult your personal physician. Neither the author nor the publisher shall be liable or responsible for any loss or damage allegedly arising as a consequence of the use or application of any information or suggestions in this book.

Because of the dynamic nature of the Internet, any World Wide Web addresses or links contained in this book may have changed since publication and may no longer be valid.

ISBN 13: 978-0-9846293-0-5
Library of Congress Control Number: 2010938688

Printed in the United States of America

Stay up-to-date with TensCam innovations. Visit *www.tenscam.com* and sign up to receive our free monthly e-newsletter, *TensCam in Action.*

Dedication

To my wonderful wife, Carolyn, and our three children,
Wendy Vranus, Michelle Byrd, and Leslie Crosby,
also Paul Sayre, Nelson Conrad Dove and
to all those who continue to support the evolution
of healthcare in the twenty-first century.

Foreword

I suppose it is only human nature that in an affluent society we should be looking to technology for medical breakthroughs to cure those remaining diseases that (so annoyingly) continue to plague us. After all, most of the diseases to which our ancestors succumbed only 100 years ago (including tuberculosis, diabetes, and polio) have been defeated, and when our hips, knees, heart valves, corneas, and other body parts wear out, they can be replaced with artificial ones.

So researchers around the world are working diligently to find cures (or at least treatments) for cancer, degenerative diseases, autoimmune diseases, chronic pain, allergies, and other challenging conditions. Scientific information accumulates, conferences are held, and optimistic announcements spout forth in the media, yet true medical breakthroughs are still rare.

Perhaps that is because real breakthroughs are not based simply upon the accumulation of knowledge. Rather, a truly original thought must come to mind and then be acted upon. A classic example would be Alexander Fleming's observation of the fungus *Penicillium* inhibiting the growth of bacteria in a petri dish.

Dr. Charles Crosby's invention, the TensCam (a CAMS device), is a product of this type of thinking. Dr. Crosby's genius was to marry the established idea that a quartz crystal can be a tool in the hand of a "healer" with another known concept: that the atmosphere's natural resonance (the Schumann resonance) is an organizing principle for the biological rhythms of the body.

It has been my privilege to observe Dr. Crosby's ideas develop, to test his inventions in my own practice, and to see him slowly collect a body of scientific knowledge explaining how the TensCam works. On this journey, he has first conceptualized the medical breakthrough and *then* uncovered the scientific explanations of how it works—quite the opposite of the vast majority of current medical research.

For the past 10 years, I have been using one of Dr. Crosby's TensCam units in my practice almost every day, several times a day. His first prototype was a simple palm-size quartz crystal with a copper wire wound around it, held together with duct tape and connected to a Liss cranial stimulator. The cranial stimulator provided the 8-Hz (Schumann) frequency that seems to be so important in the TensCam's effectiveness.

Dr. Crosby is an orthopedic surgeon as well as a skilled osteopathic physician, and the first target of his invention was somatic dysfunction, for which the TensCam works as well as or better than osteopathic manipulation in most instances. It was only while we were "playing around" with his invention that he suggested we try it on an interference field. To my (and I think his) amazement, it worked immediately, and was just as effective as the more conventional procaine injections.

Over the years, I began to use the TensCam more and more, gradually replacing my previous use of procaine injections in neural therapy. Applying the TensCam was of particular advantage in treating ganglia of the autonomic nervous system, since injections into these ganglia are not entirely safe and often require the use of 3-inch needles, stressful for both physician and patient. Eventually, I was using injections only for treating interference fields of the teeth, a situation where the combination of a homeopathic agent and procaine is often more effective than procaine alone. But a trial of the TensCam seemed warranted, and again, to my surprise, directing the TensCam "beam" through an ampoule of homeopathic held over the tooth was just as effective as injecting the homeopathic and procaine into the tissue surrounding the teeth.

Now, in this book, Dr. Crosby presents the record of his search for the scientific underpinnings of the TensCam's success. The first half of the book contains an overview of "energetics," with discussions of the human energy field, the earth's energy field, the physics of crystals, scalar energy, and the phenomenology of intention. He then introduces the concept of "interference fields," an idea taken from the German medical discipline of neural therapy. However, he uses a definition of these fields somewhat different from the classical one. Rather than being based on a disturbance of electrophysiology, Dr. Crosby's definition of interference fields is grounded in energetic terms such as "anticoherence" and "a disturbance in the human energy field."

This perspective is a helpful advance in our understanding of human functional pathology. Some in the osteopathic profession are attempting to incorporate this type of thinking into osteopathic medicine as well. Dr. Crosby quotes Fulford's prediction that "Quartz crystals will be the key to medicine in the 21st century." This prediction may prove to be correct; but understanding energetics and its effect on human health will be essential if this key is to unlock anything.

In the second half of his book, Dr. Crosby describes the application of the TensCam to clinical problems. This is the TensCam's proving ground, as no convenient way yet exists to measure the quantitative or qualitative output of scalar energy. Nevertheless, its effects can be measured, and data are steadily being accumulated by Dr. Crosby and the other physicians who use the TensCam.

Dr. Crosby presents numerous case histories demonstrating the range of conditions for which the TensCam is effective. Most of these cases are short, simple, and to the point; the TensCam shines in these applications. However, it also stands out in more complex situations that do not lend themselves well to the "teaching moments" of short histories. In these more difficult cases, the physician's skill and experience become important. Knowledge of anatomy, physiology, psychosomatics, acupuncture meridians, and other factors come into play and are essential in realizing the TensCam's full potential.

Dr. Crosby discusses some of these factors, recognition of which is necessary in order to detect interference fields and to achieve lasting responses to TensCam treatment. However, it must be remembered that the variety of presentations of functional pathology in medicine is infinite. As instructive as these examples and explanations are, the physician must be prepared to encounter and to recognize causes for illnesses that were previously unimaginable.

Medical energetics is not new, even in Western culture. For example, homeopathy was developed by the German physician Hahnemann in the early 1800s. However, it has until recently remained "under the radar" within mainstream medicine, present in everyday practice but rarely acknowledged. The dreaded placebo effect is still the bane of researchers, but the unrecognized and underappreciated companion of all skilled clinicians.

Over the past 50 years or so, many electronic devices have been invented for therapeutic purposes. Typically, these machines emit electromagnetic, thermal, or visible light energy. Dr. Crosby's invention is (to my knowledge) the first to emit scalar energy, the energy that issues from the hands of healers, the organizing energy of biology, and the energy that promises to revolutionize medical care in the 21st century.

Robert F. Kidd, MD, CM
Author of *Neural Therapy: Applied Neurophysiology and Other Topics*

Contents

Introduction

The invention of the CAMS (Crosby Advanced Medical Systems) technology resulted from a series of events, rather than from a single idea. Each event was like one dot in a dot-to-dot picture—perhaps trivial on its own, its true significance being revealed only when viewed in the context of a collection of dots, with progression from one dot to the next to create the full picture.

The series of events began in 1997 when I attended the final lecture given by the late Robert Fulford, DO (1905–1997). Dr. Fulford was a renowned osteopathic doctor and frequent popular lecturer for the Cranial Academy and the American Academy of Osteopathy. Both he and his audience were aware that the lecture he was giving would likely be his last opportunity to share his insights before he passed away. During his address, he encouraged osteopathic physicians to remember the energetic approach to health. He suggested that the energetic body should be as much a part of any diagnosis and treatment as the physical body. He also demonstrated the use of crystals as medical tools and predicted, "Quartz crystals will be the key to medicine in the 21st century."

Although parts of his lecture were received by some with a level of skepticism, Dr. Fulford's demonstration with crystals, including treatment of a medical doctor from the audience, was intriguing. The effects of the treatment appeared quite dramatic. However, because I had no background and no particular interest in crystals, I made no attempt to follow up on his comments.

At about the same time, I was introduced to Dr. Saul Liss, co-creator of the LISS Cranial Electrical Stimulator (CES), a device sanctioned by the Food and Drug Administration. The LISS CES generates microcurrents of electricity to stimulate the brain and thus to effectively treat depression, anxiety, insomnia, stress, and pain. It acts by quickly increasing the production of several neurotransmitters and other brain chemicals. The development of the CES led to the initiation of many similar technologies that fall into the class of transcutaneous electrical nerve stimulation (TENS) devices, now widely used for the relief of pain. Dr. Liss and I became friends, and our friendship offered me the opportunity to explore the deeper significance and physiological benefits of electrical stimulation.

Several months after Dr. Fulford's lecture, a patient presented me with a large quartz crystal as a gift. I thought it was interesting and quite beautiful, but I attached no particular significance to it. Thanking the patient, I placed it on a bookshelf in my office. The crystal sat there, gathering dust, for several months. Then, I injured my wrist while treating a patient. Six weeks later, there had been no improvement, and the wrist was still quite painful. Because I am an osteopathic physician who uses a variety of soft-tissue and structural alignment techniques, the injury was limiting my ability to fully treat patients, and it began to cause concern. In a moment of reflection, I looked over at the bookshelf and saw the crystal sitting there. At the same time, I remembered Dr. Fulford's comment about crystals being the key to medicine in the future. I decided to retrieve and review my notes from his lecture.

Having reviewed the notes from Dr. Fulford's lecture, I decided I had nothing to lose. I took the crystal and followed his instructions. Pointing it at my wrist, and forming the intention to heal in my mind, I took a deep breath and exhaled forcibly through my nose. The effect was almost instantaneous. I set the crystal down and began to move my wrist—slowly at first and then with increasing vigor. There was no pain—absolutely none. I was stunned by the complete freedom of movement that I was experiencing.

In that moment, I knew that more than my wrist had changed. My perception of the forces for healing that are available in nature opened before me, and my mind starting moving at a hundred miles an hour as I considered the possibilities. I wasn't sure what had happened, and I had no idea how to use the information with which I had just connected. But I knew that I needed to understand it from a scientific perspective. I knew that without an understanding of the science—the physics—I would never be able to convince my peers or even myself of its validity. In fact, this was quite possibly the reason I had given little heed to Dr. Fulford's original demonstration—it had lacked credible science to back it up. Now, I was determined to seek out and understand the underlying science.

I began to study Dr. Fulford's work in greater depth and to follow some of the same paths he had pursued. Dr. Fulford was highly influenced by Dr. Andrew T. Still (1828–1917), the originator of the osteopathic concept. I reread Dr. Still's work, paying greater attention to his emphasis on what he called the "etheric body" (another term for the energetic body) and his discussion of energy sinks, disturbances in the normal flow of energy. Dr. Still also spoke of releasing shock from the body and of a "flip in the tissue."[1(p5)] Each of these concepts came to have increasing significance as I sought to uncover the science and the mode of action in the healing I had experienced.

Dr. Fulford had also been influenced by Dr. William Sutherland (1853–1954), the developer and promoter of cranial osteopathy. In his studies with Dr. Sutherland, Dr. Fulford had come to appreciate the significance of cranial motion and its relationship to normal physiological processes. He had gained an understanding of the importance of balanced, rhythmic interchanges within the healthy body. Dr. Fulford eventually incorporated what he called a "percussion vibrator" to help re-establish normal rhythmic flow in areas of tissue disturbance. As I reread Dr. Sutherland's work, I became aware of the significance of cellular disturbances and of their impact on rhythmic interchange.

Dr. Rollin Becker (1910–1996), who had followed in the footsteps of Dr. Still and Dr. Sutherland, was another source of inspiration for Dr. Fulford. As the author of *Life in Motion* and *The Stillness of Life*, Dr. Becker had established the impact of trauma on health. He identified the impact of energy trapped by trauma and held by the body and of its effects on critical motions in the fluids of the body. Dr. Becker taught his students that it was possible for the practitioner to *feel* a patient's patterns of trauma—another concept that came to have greater significance in the work I was approaching.

Dr. Fulford was also influenced by Dr. Harold Burr and his hypothesis of an organizing energy field. In addition, Dr. Fulford had discovered the more recent research of Dr. Valerie Hunt at the University of California, Los Angeles, which essentially validated Dr. Burr's hypothesis. On the basis of the research by Dr. Burr and Dr. Hunt, Dr. Fulford became convinced that health and sickness could be defined in terms of energy, rather than entirely in terms of chemical changes in the physical body.

During my review of Dr. Fulford's research and the work of those who influenced his thinking, I began to discover many factors in nature that play a role in human health. For example, I found substantial documentation supporting the role of the Earth's background frequency as a "tuning fork for life." I also found information on scalar energy, much of which has been sequestered and kept from the public eye, although it has been a top military priority for decades. According to the research, scalar energy is connected in many ways with the organizing forces of the universe. Finally, I found that quartz crystals were nature's transducers, capable of converting electromagnetic energy to scalar energy. My task became to find a way to put all the pieces together for the creation of a healing tool for the 21st century, perhaps in partial fulfillment of Dr. Fulford's prediction.

As I paid more attention to crystals, I discovered that even though they were all made of the same substance (silicon dioxide), they resonated at different frequencies, according to the circumstances and conditions under which they had developed. Of import to my work, a small percentage of quartz crystals resonated at the frequency of the Earth and thus had the ability to naturally transfer the Earth's healing energy to people. I learned how to identify these unique crystals and how to use modern methods to amplify their effects.

More specifically, I discovered that the LISS CES, a TENS unit, emitted a frequency very close to the Earth's background frequency. A few modifications allowed it to become the FDA-sanctioned power source behind the CAMS technology. The first unit that I built was called the TensCam: a transcutaneous electrical nerve stimulator (TENS) created by Crosby Advanced Medical Systems (CAMS).

One other highly significant event as I connected the dots to create the CAMS picture came as a result of a re-injury. Years earlier, I had torn the lateral meniscus of my left knee. It occasionally bothered me, but not enough to cause serious concern, and I had continued to run regularly throughout most of my adult life. However, one day while I was running, my knee completely locked up. I could barely walk, let alone run. I considered this an opportunity to test a prototype of the device that I had recently completed. I turned on the unit, focused on healing my knee, pulsed my breath, and, treated my knee for about 3 minutes. As with my wrist, I experienced complete relief of symptoms. I *immediately* ran 5 km (3.1 miles), with no pain and no other complications, and my knee has required no further treatment in over 10 years.

The healing of my knee was the clincher—the dot that told me the picture was near completion. All that remained were the details. It took another 6 years for me to really comprehend how it all worked, but eventually the dot-to-dot picture took on dimension and color as I satisfied my own need to understand each of its parts and to know that there was sound scientific basis for each one.

This book is an attempt to present the science of scalar energy and quartz crystal technology to medical professionals and interested nonprofessionals. It is not intended to be exhaustive—quite the contrary. It is meant to include just enough information to allow the reader to understand the basis of how the CAMS technology works without being overly technical or laborious. For those who are interested in more detail, every effort has been made to include up-to-date references, including Internet references for easy access to more information. The book reflects my own healing journey and presents many of my own insights.

Part I outlines each of the pieces that play a role in the CAMS technology—everything from developing an understanding of the human energy field and its role in health to comprehending the significance of the Earth's energy field; from gaining an appreciation of the nature, organization, and properties of crystals to showing the power of intention and its connection with scalar energy. Part I also includes a discussion of what happens, at the physical level, to the tissues and fluids of the body when the CAMS device is used for treatment. In addition, it includes information on how to use the device to maximize its potential.

Part II presents an overview of the many areas where the CAMS technology can be successfully applied, with numerous examples of successful treatment.

Reference

1. Comeaux Z. *Robert Fulford, D.O. and the philosopher physician*. Eastland Press; 2002. Citing Fulford R. *Dr. Fulford's touch of life*. Pocket Books; 1996. p. 22.

Part I

Elements of the CAMS Technology

The Crosby Advanced Medical Systems (CAMS) technology brings together the best of what nature has to offer and uses it to accomplish what nature does best—balance and harmonize her creations. When balanced, the human body is capable of amazing feats, including miraculous and sometimes spontaneous healing. The elements of the CAMS technology are all found in nature. They include the planet Earth itself, or rather the resonant frequency of the Earth, which has a balancing effect on all its creatures. The technology also incorporates the Earth's most abundant mineral, quartz. This mineral has played a key role in ushering in our technological age. CAMS technology also uses scalar energy, the underlying force of the universe, which we are just beginning to understand. Together with conscious intention, these elements combine to provide organized, healing support for the human energy field. Each will be discussed in depth in the chapters of Part I as the foundation of CAMS technology is laid out and we explore the science behind this approach to healing.

Chapter 1

The Human Energy Field

For years, traditional scientists have viewed the human body as solid matter, composed of particles that combine in ever-increasing numbers to form atoms, cells, tissues, and organs, all surrounded by a protective skin. Today, we are beginning to understand that all living matter is in fact made of energy and that on every level, the human body is composed of multiple, interacting energy fields. Even though many of these energy fields have been difficult to detect and to quantify, mounting evidence suggests that life itself can be defined in terms of the flow of energy.[1] Every living thing resides within one or more energetic fields. Likewise, every living thing radiates life-force energy, which has been referred to by some as "subtle energy."

Valerie Hunt, PhD, a neurophysiologist at the University of California, Los Angeles (UCLA) and author of the book *Infinite Mind: Science of the Human Vibrations of Consciousness*,[2] was one of the first to verify the subtle energy field that surrounds the human body. In the 1970s, she authenticated what healers have understood for centuries: the existence of a dynamic bioenergetic field around all living things. After connecting volunteers to wireless biofeedback equipment that had been used in research conducted by the National Aeronautics and Space Administration (NASA), Hunt monitored the volunteers' biodata and compared it to descriptions of the colors and patterns perceived by successful healers and respected aura readers.

Hunt found that a person's biodata would appear on the monitor at the exact moment aura readers observed changes in the energy field around that person. Hunt's ground-breaking work opened the door for more extensive research and led her to conclude, "I can no longer consider the body as organic systems or tissues. The healthy body is a flowing, interactive, electrodynamic energy field."[2(p48)]

Hunt eventually discovered that the composite bioenergy field of a healthy human being is composed of balanced, coherent energy patterns flowing across the full spectrum of frequencies. In her work, coherence showed up on graphs as smooth, gentle waves that were evenly distributed throughout the frequency spectrum. Lack of coherence was exemplified by irregular, broad wavebands distributed erratically across graphs.[2(p328–348)]

Coherence

In physics, coherence is defined as a fixed-phase relationship between waves.[3] Coherence can also be understood as a measure of order or organization, rather than chaos. When the components of a system are coherent, they are organized in such a way that they can sustain the ongoing flow of energy and information. For an organism, coherence also reflects the ability to function as a whole, rather than as independent parts.

Hunt discovered a distinct relationship between the coherence of a person's bioenergy field and his or her health. She determined that disease had its roots in what she called "anticoherent patterns" in the energy field. Describing this insight, she wrote, "All diseases are caused by a break in the flow, or a disturbance in the human energy field. Eventually, this disturbance is transferred to the organ system creating functional and ultimately, destructive changes."[2(p76)] According to years of research conducted by Hunt, anticoherence (holes, tears, chaos, or other disturbances) in the bioenergy field precedes sickness and ill health. Even more importantly, correction of such disturbances results in the disappearance of symptoms.[2(p240–244)] Hunt devoted many years to creating maps of the human bioenergy field and to determining the relationship of the energetic field to health, emotional stability, spiritual development, and human behavior.

The bioenergy field as template lines

In the 1930s, Yale University School of Medicine neuroanatomist Harold Burr, MD, theorized that bioenergy fields were the "organizing principle" that kept living tissue from falling into chaos. He noted that an electrodynamic field could be detected in all early embryos, as well as in plants and animals without neural or perineural tissues.[4] Burr referred to these fields as "L-fields" (meaning "life fields"), and he was one of the first to measure the changes that occurred in these fields during sleep, growth, drug usage, and tumor formation. He even measured the L-fields of trees and monitored their responses to light, moisture, weather, and the phases of the moon.

Burr determined that developing organisms were destined to follow a prescribed growth pattern generated by their respective L-fields. In his well-known book, *The Fields of Life*, he discussed his years of work and many of his findings.[5] Hunt continued Burr's work, demonstrating that acupuncture meridians and connective tissue worked cooperatively to provide the doorway to and the channels for the flow of energy and information from the bioenergy field to and throughout the physical body.[2(p242)]

According to traditional theory, the acupuncture system is a circulatory system for mobilizing energy and for intercommunication throughout the body. This system is formed *before* the tissues themselves develop.[6(p51–52)] Although still physically unverifiable, acupuncture points are known to possess properties different from their surrounding tissues. For example, their resistance is lower, by a factor between 10 and 100, than that of the surrounding skin.[7,8] Energy follows the pathway of least resistance, and acupuncture points thus provide a conduit for the conduction of energy from the bioenergy field to the physical body. Similarly, healthy connective tissue forms a conductive matrix that links all cells to one another. It is composed mostly of water and is therefore liquid crystalline in nature and hence able to intricately link the liquid crystalline components of the body (collagen, cell membranes, DNA, muscle fibers, etc.).[8] The liquid crystalline properties of the connective tissue matrix create an excitable electrical continuum for rapid intercommunication throughout the body.[8]

Robert Becker, an orthopedic surgeon and coauthor of the widely acclaimed book *The Body Electric*, made this same connection in the 1970s during his work on limb regeneration. He proposed that connective tissue was semiconductive, a property that is possible only in materials having an orderly molecular structure.[9(p92–94)] He also established the role of electrical currents during the healing process and implied the presence of a larger organizing field.

The process of Kirlian photography, invented in Russia, provides further evidence for the bioenergy field as a template or organizing field. Early Kirlian photographs revealed the energy field around leaves. Surprisingly, in photographs taken after the leaves were cut into pieces, the energy field of the *whole* leaf was still depicted. This effect has been referred to as the "phantom leaf phenomenon." It is reminiscent of Burr's work with salamanders, which showed that the energy field of a young salamander resembled that of the adult.[5] However, the most striking phenomenon demonstrated by early Kirlian photography was revealed by the Romanian researcher Dumitrescu, who added a new twist to the phantom leaf phenomenon. Dumitrescu cut a circular hole in a leaf before taking a Kirlian photograph of it. The image of the leaf displayed another smaller leaf inside the hole, and this smaller leaf also had a hole in its center.[10(p62)] This is exactly what occurs with a hologram. If a hole is cut in the middle of a hologram, and the resulting cut-out is held up to a laser light, a tiny intact object is visible in the film piece. The work of Dumitrescu is strong evidence not only for the existence of an organizing bioenergetic field, but also for its holographic nature.

The bioenergy field is, in all likelihood, at least similar to a hologram. Such a similarity would be in harmony with the Nobel prize–winning work of Karl Pribram, PhD, who theorized that the brain itself is a hologram. His work explained why full memory persists among some individuals who have had parts of their brains destroyed.[11] The idea of a holographic brain memory provides the basis for the holographic nature of the bioenergy field itself, whereby the field is a blueprint containing information that supports cellular activity and the ongoing health of the entire body. When the blueprint becomes distorted, cellular health, as well as psychological health, can be affected.

Both Burr and Hunt were instrumental in establishing that changes in the bioenergetic field preempted changes in the physical body. Ultimately, a disturbance in the energy field produces a disturbance in the tissues of the body. Sometimes, the effects are immediate, as is the case with accidents and pain. (According to Hunt, pain occurs in the bioenergy field before it is felt in the body.[2[p244]]) Other times, a disturbance in the energy field can take years to manifest as physical symptoms, as is the case with some early childhood traumas and emotional issues. Regardless of how long it takes for the symptoms to manifest, the re-establishment of coherence within the bioenergy field often results in rapid reduction or even complete resolution of the symptoms.

Initiation of healing in the bioenergy field

Cells grow old and die. With the exception of brain cells, all of the body's cells are replaced on a continuing basis. The only explanation for how new cells are created and organized in the pattern of the cells they replace lies in the existence of a template—a bioenergy field. Burr said of this phenomenon:

> *When we meet a friend we have not seen for six months there is not one molecule in his face which was there when we last saw*

> *him. But thanks to his controlling L-field, the new molecules have fallen into the old familiar pattern and we can recognize his face.*[12]

Thus, health and healing are both functions of a dynamic, coherent energy field. They are, in fact, initiated there.

Barbara Ann Brennan, former NASA astrophysicist and author of the well-known book *Hands of Light*,[13] gave early credibility to the concept of energy healing and brought it into the realm of rigorous science. During an interview, she described how her work with the energy field initiated the healing response:

> *A healing ... would usually begin at the feet and work up the body, charging and balancing the field. I would go to the specific areas of the body that needed work and clear, charge, balance, and restructure the field. Restructuring involves rebuilding the specific pattern of the field in the area that's been distorted.*[14]

Experiments conducted by Hunt also substantiated the role of the bioenergetic field in initiating the healing process. For example, when she simultaneously recorded the frequency pattern of a dysrhythmic heart and the pattern in the bioenergetic field, the two recordings revealed the same anticoherent pattern. When the coherence in the energy field was improved, the anticoherent pattern in the heart also dissipated. At the same time, the patient's breathing was normalized, while the heart itself came under automatic control.[2(p247)] My colleagues and I have often seen a similar, almost instantaneous, reversal of symptoms when using the CAMS devices, which operate by bringing coherence to areas of the bioenergy field. Many examples of such occurrences are presented in Part II of this volume, which includes case studies and a discussion of potential applications of the CAMS technology.

Thoughts and emotions

After years of research, Hunt became a firm believer that thoughts, feelings, and unresolved emotional conflicts were reflected in the human energy field. Through her work, traumas and repetitive negative thoughts (the result of unsupportive beliefs) were revealed as anti-coherent areas in the bioenergy field.[2(p95)]

Not only does the human energy field house the physical body, providing information for growth, development, and healing, it is also a carrier of the patterns that govern emotional, mental, and spiritual stability. It is a repository of information that reflects a person's past experiences (especially traumas), beliefs, and emotional and spiritual state. Unexpressed emotions, as well as what the person regularly thinks and feels, reside as patterns and/or disturbances in the energy field. Eventually, they create the person's reality. All this gives validity to the connection between body, mind, and spirit. This connection is found in the bioenergy field.

For thousands of years, philosophers have pointed out that our thoughts determine our experience of the world. These original thinkers understood that thoughts, beliefs, and certainly emotions have the power to enslave or to enlighten. In 1997, Candace Pert released the revolutionary book *Molecules of Emotion: Why You Feel the Way You Feel,*[15] summarizing years of work and bringing substance to the phrase "thoughts are things." Pert sealed the connection between the physical body, the emotions, and the human energy field. Her work and her book revealed that neuropeptides—electrochemical signals—are activated by thoughts and emotions and that they change the chemistry and the electricity of every cell in the body.[15] Pert demonstrated the sequence of events initiated by thoughts and emotion, which were followed by changes in chemistry, changes in physiology, and ultimately disease. In one interview, she stated:

> *There is data that shows that there are fields around the body, and that energy is emitted from a healer's hands. There is a lot*

> *of data showing that when you assume an emotional state, this somehow aligns your internal chemistry and in turn your internal vibrations. The term vibration is actually very accurate because the cellular receptors are not just little "lock[s] and keys" they are actually vibrating as they pump ions and information through the cell membrane. A little pain in your back could be connected to an emotion, a memory, or a problem elsewhere.*[16]

Many healthcare professionals understand the body–mind–spirit connection with respect to memories and emotions. Certainly, bodyworkers are aware of the somatic or cellular memory—a person's ability to store memories and other information within the cells and tissues. It is not uncommon for tissue manipulation during massage and other forms of hands-on therapy to bring up old memories for resolution and release. What practitioners are now beginning to recognize is that memory is also held in the bioenergetic field. In fact, bodywork may be the stimulus that draws memories from the energy field into conscious awareness. From one perspective, thoughts and emotions can be considered the "currency of exchange" between spirit, mind, and body.

Many practitioners are also aware that most physical problems have an emotional component. Emotional pain can take many forms: grief, guilt, regret, pent-up anger, worry, depression, a feeling of being stuck or trapped, heartbreak, fear, and shame. Individuals may carry such emotional pain and be unaware of it at the conscious level. However, releasing the emotional component stored in the energy field (by restoring coherence) often provides a bridge to awareness and/or the resolution of numerous physical problems. In our clinical work with the CAMS technology, my colleagues and I have found that restoring coherence to anticoherent areas of the energy field (caused by unresolved emotional conflict) resolves a high percentage of physical problems.

In the next chapter we will consider the Earth's energy field and its influence on health and healing.

References

1. Ho MW. *The rainbow and the worm: the physics of organisms*. 2nd ed. World Scientific; 1998.
2. Hunt V. *Infinite mind: science of the human vibrations of consciousness*. Malibu Publishing Co.; 1996.
3. *Random House Webster's college dictionary*. 2nd ed. Random House; 1997.
4. Burr H, Northrup F. The electro-dynamic theory of life. *Q Rev Biol*. 1935;10:322–333.
5. Burr H. *The fields of life*. Ballantine Books; 1972.
6. Gerber R. *Vibrational medicine*. HarperCollins; 2001.
7. Reichmanis M, Marino AA, Becker RO. D.C. skin conductance variation at acupuncture loci. *Am J Chin Med*. 1976;4(1):69–72.
8. Ho MW, Knight DP. The acupuncture system and the liquid crystalline collagen fibres of the connective tissues: liquid crystalline meridians. Institute of Science in Society. Available from: *http://www.i-sis.org.uk/lcm.php*
9. Becker R, Selden G. *The body electric*. William Morrow and Company; 1985.
10. Gerber R. Future trends in healing. In: Marberry SO, editor. *Innovations in healthcare design: establishing a new paradigm. Selected presentations from the first five Symposia on Healthcare Design*. John Wiley & Sons, Inc.; 1995. p. 49–69.
11. Prideaux J. Comparison between Karl Pribram's "holographic brain theory" and more conventional models of neuronal computation. American Computer Science Association. Available from: *http://www.acsa2000.net/bcngroup/jponkp/*
12. Becker R, Selden G. *The body electric*. William Morrow and Company; 1985. Quoting H. Burr.
13. Brennan BA. *Hands of light: a guide to healing through the human energy field*. Bantam; 1988.

14. Windsor R. Healing with the human energy field. An interview conducted with Barbara Ann Brennan. In: *Spirit of Ma'at online magazine*. Spirit of Ma'at LLC. Available from: *http://www.spiritofmaat.com/archive/nov1/brennan.htm*

15. Pert C. *Molecules of emotion: why you feel the way you feel.* Scribner; 1997.

16. Millett R. Interview with Candace Pert: healing emotions. Available from: *http://www.juliajohnson.co.uk/pages/articles/interview-with-candace-pert.php*

Chapter 2

The Schumann Resonance: The Earth's Energy Field

The Earth itself is housed within a huge energy field, the result of standing waves formed between the positively charged ionosphere and the negatively charged surface of the Earth.[1(p98)] The Earth's energy field is activated and sustained by lightning (about 5,000 strikes per minute around the planet), by solar radiation, and by heat from the core of the planet.[2(p6)] The presence of the Earth's energy field was first identified and quantified by Nikola Tesla (1856–1943), who found that the Earth was "literally alive with electrical vibrations."[3(p222)] Much of Tesla's work was based on this and other information he gleaned about the Earth's energy field, although it was ignored by many of those in the scientific community during his lifetime.

Tesla was the first to utilize the ionosphere for the transmission of radio waves. More than half a century later, in 1954, the German physicist W.O. Schumann recalculated the frequency of the Earth's energy field, reporting it as 7.83 Hz (1 hertz = 1 cycle per second).[4(p183)] Schumann was given credit for this discovery, which was named in his honor: the Schumann resonance.

Just as Tesla's identification of the Earth's resonance was ignored, Schumann's discovery might also have gone unnoticed had it not been for a physician named F. Ankermüller, who read Schumann's work. Ankermüller made the connection between the Schumann resonance and human brain waves. This correlation was later verified by one of Schumann's graduate students, who compared human electroencephalographic (EEG) recordings with the frequencies emitted by the Earth.[5(p25)] These findings opened a new vista to understanding the connection between human beings and the Earth's energy field.

Today, we recognize that the Schumann resonance is essentially a "tuning fork" for all life. It acts as a background frequency, influencing a variety of biological functions. At the very minimum, it synchronizes brain waves, balances circadian rhythms, improves performance, and plays a role in hormone balance.[1(p98–101)] As many have noted, its influence is far-reaching,[1(p109),6(p67)] and it has the potential to bring coherence back to an anticoherent human energy field.

Effects of the Schumann resonance

Since Schumann's early work, others have investigated the synchronizing effects of the Schumann resonance on the human brain. For example, Neil Cherry, one of the first researchers to make the connection between cell phone radiation and many health problems, determined that the Schumann resonance provided a frequency range that synchronized and continuously stabilized the brain.[7]

Cherry also uncovered a correlation between the Schumann resonance and the melatonin–serotonin cycle that determines sleep–wake patterns.[8] To him, it was obvious that the Schumann resonance also had an effect on many organs with functions that cycled between night and day.

R. Wever, from the prestigious Max Planck Institute in Germany, demonstrated the influence of the Schumann resonance on circadian rhythms. He began by building a shielded, underground environment to screen out the Earth's electromagnetic field. Volunteers who lived in this environment for extended periods of time experienced desynchronization of their physiological rhythms and an assortment of physical and emotional symptoms. When the volunteers were subsequently exposed to an artificial Earth field, their rhythms stabilized.[9]

Some of the symptoms experienced during Wever's studies were reported by the first astronauts and cosmonauts,[10] who were not exposed to the Earth's Schumann resonance while they were in space. Modern spacecraft are said to contain a device that simulates the Schumann resonance to compensate for the absence of the Earth's field in space.

Several researchers have established the connection between the Schumann resonance and performance, particularly reaction time. When frequencies other than the Schumann resonance are exclusively supplied to the human body, reaction time is significantly slowed. Conversely, fields that closely approximate the Earth's natural resonance (8–10 Hz) have been shown to speed up reaction times.[11,12]

In similar studies in the 1970s, Valerie Hunt documented the influence of the Earth's field on performance. When electromagnetic fields were reduced or withdrawn from a test environment, participants experienced incoordination, sensory instability, and emotional distress. Measurements of each participant's personal energy field showed gross disturbances and anticoherence, but the field was completely restored when the natural electromagnetic environment was re-established.[6(p30–32)]

Schumann harmonics

The Schumann resonance (accepted as having a frequency of 7.83 Hz) is actually an average of the Earth's fundamental frequency, which varies both daily and seasonally. This frequency and its strength depend on the distribution of global thunderstorm activity, sunspots, geomagnetic storms, eclipses, and the positions of the moon and other planets.[1(p99)] Over time, the frequency of the Earth has been measured at between 1 and 40 Hz. By way of comparison, audible sound is in the frequency range of 20 to 20,000 Hz. At times, the Schumann resonance can actually be heard by those with an acute sense of hearing in the lower frequency range. It has been described as a "low hum."

During thunderstorms, the thunder itself interfaces with the Schumann resonance and generates harmonics or overtones. In reality, the Schumann resonance is not a single frequency. Rather, it gives rise to an entire frequency chord, the Schumann resonances. The lowest and most intense frequency is 7.83 Hz, with higher harmonics occurring at 16, 32, and 48 Hz. This frequency range closely matches the electromagnetic spectrum of the human brain.[1(p94)] This is what caught Ankermüller's attention when he read Schumann's original work.

The basic frequency of the Schumann resonance (7.83 Hz) is the foundation from which supportive harmonics and overtones are generated in the Earth's field. Each of these overtones no doubt plays a role, balancing the planet and supporting its life forms. The same is true when it comes to the human body. Supplying the basic Schumann frequency opens the door to the entire range of supportive frequencies that sustain health and well-being. As such, the Schumann resonance is the foundation for healing.

The Schumann resonance for healing

Although the existence of the Schumann resonance is an established scientific fact, very few are aware of its importance for the maintenance of health or its usefulness in supporting the healing process. Simply standing barefoot on the surface of the Earth has now been documented to provide a variety of physiological and emotional benefits.[13] This position literally grounds the human body and provides electrical stabilization of the human energy field. This phenomenon is now being intently studied in light of the fact that modern life is characterized by numerous electromagnetic interference patterns that affect the human energy field.[2]

However, simply reconnecting with the Earth and its resonant frequency is not enough to overcome localized disturbances (specific areas of anticoherence) in the human energy field, especially if those disturbances have existed for a long time. Isolated areas of anticoherence in the human energy field, which have been referred to as "interference fields," are caused by trauma, either physical, mental, or emotional. These interference fields can produce long-standing disturbances in the electrochemical function of tissues, as well as destabilization of the autonomic nervous system.[14(p24)] Interference fields, discussed in greater detail in Chapter 6, *do* respond to the Schumann resonance, but a more focused approach is required.

Focusing the Schumann resonance

In the 1970s, Robert C. Beck began research on the brain wave activity of healers. Gathering a number of practitioners who practiced a variety of forms of healing, he recorded brain wave frequencies while they performed their work. He discovered that all of the healers registered brain waves averaging 7.8–8.0 Hz. Subsequent studies determined that during "healing moments," the brain waves of the healers were synchronized, in terms of both phase and frequency, with the Earth's Schumann resonance.[15]

Not only do the brain waves of successful healers become synchronized with the Schumann resonance, but these practitioners are also able to amplify and focus the frequency of the Schumann resonance. This explains how hands-on energy work has sometimes been able to accomplish what other methods could not. According to James Oschman, respected physiologist and author of the book *Energy Medicine: The Scientific Basis*, the electrical currents from a healer's brain waves are conducted through the perineural and vascular systems.[1(p104)] They can be amplified by as much as 1000 times and delivered to very specific areas of the human energy field.[1(p104)] The ability to amplify and to focus the Schumann resonance is a key to re-establishing coherence to interference fields. It is, quite possibly, the reason why some healers are so successful. Amplification and focusing of the Earth's Schumann resonance are major factors in the design and function of the CAMS technology.

Interestingly, dolphins, crickets, and other life-forms are said to produce sound waves that coincide with the Schumann resonance. Research has shown that human contact with dolphins induces both the alpha brain wave state and hemispheric synchronization in the brain.[16] This could be one reason why many people report therapeutic benefits from contact with dolphins. Dolphins communicate by generating biosonar sound energy (echolocation), which couples with the human body through water. The enhanced efficiency of water to carry sound (nearly five times greater than air) may effectively amplify the Schumann resonance and deliver its healing effects.

Crystals also have exceptional wave-controlling properties, and they exhibit the ability to amplify and focus energetic information. The next chapter discusses how and why crystals work in this capacity.

References

1. Oschman J. *Energy medicine: the scientific basis.* Churchill Livingstone; 2000.
2. Ober C, Sinatra S, Zucker M. *Earthing: the most important health discovery ever?* Basic Health Publications; 2010.
3. Tesla N. Communicated to the Thirteenth Anniversary Number of the *Electrical World and Engineer*, March 5, 1904. In: *The fantastic inventions of Nikola Tesla.* Adventures Unlimited Press; 2009.
4. Schumann WO, König H. Über die Beobachtung von Atmospherics bei geringsten Frequenzen. *Naturwissenschaften.* 1954;41:183.
5. König HL. Bioinformation—electrophysical aspects. In: Popp FA, Becker G, König HL, Peschka W, editors. *Electromagnetic bioinformation.* Urban und Schwarzenberg; 1979.
6. Hunt V. *Infinite mind: science of the human vibrations of consciousness.* Malibu Publishing Co.; 1996.
7. Cherry N. Human intelligence: the brain, an electromagnetic system synchronised by the Schumann resonance signal. *Med Hypotheses.* 2003;60(6):843–844.

8. Cherry N. Schumann resonances, a plausible biophysical mechanism for the human health effects of solar/geomagnetic activity. *Nat Hazards.* 2002;26(3):279–331.
9. Oschman J. *Energy medicine: the scientific basis.* Churchill Livingstone; 2000. Quoting Wever R. ELF effects on human circadian rhythms. p. 101.
10. Ehlers H. What everyone should know about magneto therapy [interview with Dr. Wolfgang Ludwig]. *Explore! For the Professional.* 1997;8(1).
11. Oschman J. *Energy medicine: the scientific basis.* Churchill Livingstone; 2000. Quoting Hamer JR. Effects of low level, low frequency electric fields on human time judgement. p. 101.
12. Oschman J. *Energy medicine: the scientific basis.* Churchill Livingstone; 2000. Quoting Beatty J. Learned regulation of alpha and theta frequency activity in the human electroencephalogram. p. 101.
13. Chevalier G, Mori K, Oschman JL. The effect of earthing (grounding) on human physiology. *Eur Biol Bioelectromagn.* 2006 Jan 31;:600–621.
14. Kidd R. *Neural therapy: applied neurophysiology and other topics.* Custom Printers; 2005.
15. Oschman J. *Energy medicine: the scientific basis.* Churchill Livingstone; 2000. Quoting Beck R. Mood modification with ELF magnetic fields: a preliminary exploration. p. 107–108.
16. Cole D. Electroencephalographic results of human–dolphin interaction: a sonophoresis model. Available from: *http://www.aquathought.com/idatra/symposium/96/sonophor/sonophor.html*

Chapter 3

Crystals: Transduction, Amplification, and Focus

Because of their unique molecular organization and their ability to transduce and amplify information, crystals have played a pivotal role in ushering in our technological age. The wartime decision by the US military to convert its radio equipment to crystal control changed the course of history and resulted in the creation of a new industry. Virgil Bottom,[1] in his brief history of the quartz crystal industry, reported that a 1943 issue of *LIFE Magazine* carried the following excerpt from a letter to the editor, in which Gerald Holton of Harvard University outlined the importance of crystals for the developing technology:

> *These little glass-like quartz wafers are perhaps the most remarkable of all the tools science has given to war. When the story of the almost incredible progress in research and manufacture of radio crystals can be told, it will prove to be a tale of one of the war's greatest achievements. No less significant will be the fruit of these advancements to a new world at peace where crystals will be the vibrating hearts of most telecommunication equipment.*

In accordance with Holton's prediction, quartz crystals are used today for a wide variety of technological purposes. They have electrical wave-controlling properties that make them ideal for sound and light amplification in radios, televisions, and lasers. Crystals are also used as frequency control devices in clocks, computers, navigation equipment, chemical reaction monitors, biomedical sensors, telecommunications timing modules, global positioning systems, and many other applications too numerous to list.

Crystals are a unique class of materials defined by their three-dimensional molecular structure. Quartz crystals are created when the elements silicon and oxygen combine to form a lattice based on repeating tetrahedral geometry. The repeating pattern and the resulting lattice give quartz crystals unique properties.

Piezoelectricity and semiconduction

Piezoelectricity is an example of a collective property that is based on the way in which atoms are organized in a crystal. When crystalline quartz is in its natural state, the positive and negative ions of silicon and oxygen are evenly and uniformly distributed, and the quartz is an electrically neutral, nonconductive material. However, when a quartz crystal is compressed, even slightly, electrons shift within the lattice, which creates a voltage along one axis. Crystals that produce an electrical voltage in this manner are said to be piezoelectric (capable of generating electricity from pressure). The phenomenon also works in the opposite direction: if current is sent through a crystal, a pressure effect is created, which produces mechanical vibrations. Crystals with piezoelectric properties provide a convenient way to convert one form of energy to another, in a process called "transduction." Because quartz crystals are transducers, they are able to convert mechanical pressure into electrical current and other forms of energy.[2(p152)]

Quartz crystals are also semiconductors. They possess the characteristics of insulators and at the same time can be made to conduct energetic information, as explained by their piezoelectric potential. Semiconductors can accomplish far more than simple conductors.

They are capable of *processing* energy and information, with the capacity to detect, switch, store, modulate, filter, rectify, and amplify energy.[2(p93)] These qualities of quartz crystals are widely utilized in modern technology.

Crystals as amplifiers

The ability of crystals to amplify energy is also due, in part, to their molecular organization. The crystalline lattice provides an organized matrix within which energy can flow more efficiently. Organization also generates the flow itself, as described by Nobel prize–winning biochemist Albert Szent-Györgyi:

> *If a great number of atoms be arranged with regularity in close proximity, as for example in a crystal lattice, single valency electrons cease to belong to one or two atoms only, and belong instead to the whole system. A great number of molecules may join to form energy continua, along which energy, namely excited electrons, may travel a certain distance.*[3]

In other words, some of the electrons within a crystal are free to move from molecule to molecule, conveying energy and information in the process. Within a quartz crystal, energy may be conveyed via electrons or via "holes," the spaces created as electrons move. Holes behave like positively charged particles, moving in much the same way as the electrons that have vacated them. Together, electrons and holes generate waves of energy.[2(p93)]

From another perspective, crystals are valves or gates. The tiniest amount of pressure—as little as someone picking up and holding the crystal—opens the gate, allowing energy to flow, in much the same way as opening the valve on a water faucet allows water to flow. The application of a small amount of force to open the valve releases an abundant flow of water. In this way, crystals are capable of converting a small amount of input into a powerful flow of energy.[4]

Marcel Vogel (1917–1991), an IBM research scientist who was instrumental in the development of liquid crystal technology, discovered the powerful energy that could be accumulated and conveyed using crystals. One of his early experiences with crystals occurred when a co-worker complained of back pain. Intuitively, Vogel removed a crystal from his lab coat pocket. Directing it toward his friend, he pulsed his own breath. This sent a charge through the crystal that knocked his friend to the floor, but when his friend got up, the back pain was gone.[5] This incident was an eye-opener for Vogel. It carried a profound message: "Crystals are tools that will accumulate a charge and transmit it to a designated destination via breath and intent."[6]

Vogel devoted much of the remainder of his life to gaining the wisdom to use crystals appropriately. His work helped to quantify their usefulness as tools for healing.

Crystals and water

Vogel's research into the therapeutic applications of crystals led him to investigate the relationship between quartz crystals and water. He suspected that one of the reasons that quartz crystals had such an impact on healing was because their molecular geometry was similar to that of water.[7] Like quartz crystals, water in the human body also forms a repeating tetrahedral pattern. As such, water creates a liquid crystal that is involved in the relay and storage of information, in the amplification of biological signals, and in the transduction of a variety of forms of energy.[8] Vogel reasoned that because water makes up about 75% of the physical body of all biological life-forms, crystals are the perfect mechanism for delivering energy to the physical body.[5] Vogel also felt that the human energy field was anchored to the physical body by means of water molecules.[5]

Research is now revealing that water within the human body and other organisms is very different from the water on the surface of the planet. Water within an organism is organized around other liquid crystalline components (cell membranes, DNA, proteins, connective tissue, etc.), to create a liquid crystal continuum with properties similar to those of quartz crystals. As such, water within the human body has piezoelectric and semiconductive properties—just like quartz crystals. It behaves much like a transistor and is capable of storing information, translating one form of energy into another, and participating in the almost instantaneous transmission of signals and other information throughout the body.[8]

Many scientists are beginning to understand that the entire human body operates as a living crystal. Even bones have elements of crystallinity. They are piezoelectric and are therefore responsive to electromagnetic fields.[2(p52)] This is one reason why magnetic fields have been successful in helping to heal bone fractures. Tissues, cells, and fluids of the body have a definite molecular organization that is supported by liquid crystalline water and further supported by the organization of the bioenergy field.

Crystals for storing and transferring information

Through his work, Vogel learned that the information accumulated in a crystal could be either stored or transferred.[5] In 1994, a group of physicists at Stanford University finally proved the storage capacity of crystals when they succeeded in storing and retrieving holographic images from a crystal. According to the news release from Stanford University:

> *Invisible to the viewer, the patterns of electrons inside the clear block [crystal] are rearranged and a series of images are stored. Next . . . [it] reflects the pattern of those rearranged electrons and relays them to a video camera, then to a computer. Its monitor displays a video of a bird flapping its wings, then a digitized version of the Mona Lisa. This is the first working demonstration of a technology that scientists have been talking*

> *about for 25 years: a system that can store videos, sound, and data as holograms.*[9]

I have already mentioned the unique molecular structure of quartz crystals, which is based on a tetrahedral arrangement of silicon and oxygen atoms. Although the SiO_4 tetrahedron can be considered the building block of quartz, the basic structural unit of quartz crystals is a group of three SiO_4 tetrahedrons. Each group of three tetrahedrons is connected to other groups immediately above and below it to form a vertical column along the lengthwise (c) axis of the crystal. The entire quartz crystal is made up of these chains.[10] Even more interesting is the fact that the chains themselves wind around each other to form single and double helices, reminiscent of the double helix of DNA.

Another distinctive feature of quartz crystal structure is the presence of channels that run lengthwise through the entire crystal, down the center of each helix. Bruce Lipton, PhD, author of *The Biology of Belief*, has referred to the DNA as the double helix memory disk of the cell.[11(p92)] Helices and channels may contribute to a crystal's information storage capacity, as well as to its ability to gather and direct coherent energy.

A protocol for healing

Guided by the molecular geometry of water, Vogel developed a unique way of cutting crystals.[12(p139)] In particular, he is known for the creation of a variety of crystals cut to specific patterns to allow them to resonate at specific frequencies.[6] During his years of work with crystals, Vogel found that an appropriately tuned crystal could restore coherence to anticoherent areas of bioenergy fields of both humans and plants. He developed a protocol that involved a tuned quartz crystal activated by the hand and powered by the breath and by intention. In this protocol, a tiny vibration (pressure from the hand) was converted into an electrical charge for release into the bioenergy field.

Vogel felt that the process of breathing, as a component of the healing process, involved several stages. Inhalation is for gathering energy and building life's intention; when a crystal is being used, the in-breath charges the crystal. Holding the breath, even momentarily, brings the charge to a peak. Exhalation, especially through the nose, releases the charge and the intent. In Vogel's protocol, the crystal is held in the hand and intention is directed through the crystal, where it picks up coherence and amplifies the energy.[13] Valerie Hunt also noted that crystals could make a healer's energy more focused and coherent, but she observed that by themselves, crystals had no measurable effect as a healing tool.[14(p262)]

Several years of clinical experience have shown that quartz crystals tuned to the resonance of life (the Schumann resonance) represent the perfect mechanism for delivering coherent transmissions to the bioenergy field and thus to the innermost parts of the physical body through the crystalline water medium.[15] CAMS devices use quartz crystals selected to resonate at the Schumann resonance. Research with the CAMS technology has revealed rapid yet gentle reorganization of anticoherent areas in the bioenergy field, which often brings immediate resolution of many physical and emotional symptoms.

In the next chapter, we will explore the connection between crystals and scalar energy.

References

1. Bottom V. A history of the quartz crystal industry. In: *Proceedings of the 35th Annual Frequency Control Symposium;* 1981. p. 3–12. Available from: *http://www.ieee-uffc.org/main/history.asp?file=bottom*
2. Oschman J. *Energy medicine in therapeutics and human performance.* Butterworth and Heinemann; 2003.
3. Oschman J. *Energy medicine in therapeutics and human performance.* Butterworth and Heinemann; 2003. Citing Szent-Györgyi A. The study of energy levels in biochemistyr. *Nature.* 1941;148:157–159.
4. Beaty WJ. How do transistors work? Part II. 1995. Self-published. Available from: *http://amasci.com/amateur/transis2.html*

5. The legacy of Marcel Vogel. Lifestream Associates. Available from: *http://www.vogelcrystals.net/legacy_of_marcel_vogel.htm#2*
6. Quartz crystal basics. Lifestream Associates. Available from: *http://www.vogelcrystals.net/Crystal%20Basics.htm*
7. Dr. Marcel Vogel speaking on the structuring of water [video]. Available from: *http://www.youtube.com/watch?v=dhyuhRwxMtw*
8. Ho MW, Knight DP. The acupuncture system and the liquid crystalline collagen fibres of the connective tissues: liquid crystalline meridians. Institute of Science in Society. Available from: *http://www.i-sis.org.uk/lcm.php*
9. News release [regarding digital holographic video/data storage system]. Stanford News Service; 1994 Aug 4. Available from: *http://news.stanford.edu/pr/94/940804Arc4171.html*
10. Akhavan A. The quartz page. Self-published; 2010. Available from: *http://quartzpage.de/gen_struct.html*
11. Lipton B. *The biology of belief: unleashing the power of consciousness, matter and miracles.* Mountain of Love/Elite Books; 2005.
12. Comeaux Z. *Robert Fulford, D.O. and the philosopher physician.* Eastland Press; 2002.
13. Vogel M. Using the Vogel-cut healing crystal. Lifestream Associates. Available from: *http://www.vogelcrystals.net/how%20to%20use.htm*
14. Hunt V. *Infinite mind: science of the human vibrations of consciousness.* Malibu Publishing Co.; 1996.
15. Crosby C, et al. TensCam low-back pain and scalar wave therapy: clinical study. University of Central Florida; 2002. Available from: *http://www.tenscam.com/content.cfm?id=22*

Chapter 4

Scalar Energy

In the late 1800s, Nikola Tesla, the electrical engineering genius and the man responsible for developing alternating current, discovered a type of energy that he referred to as "radiant energy." He repeatedly stated that this energy was nonhertzian and that, unlike other forms of wireless transmission, it did not dissipate with distance. In other words, the energy and information transmitted via radiant energy were unrestricted, even by the speed of light.[1] Of this energy, Tesla said:

> *Throughout space, there is energy. It is a mere question of time when men will succeed in attaching their machinery to the very wheelwork of nature. Many generations may pass but in time our machinery will be driven by a power obtainable at any point in the universe.*[2]

Despite Tesla's many contributions to science, his discovery of radiant energy was initially ignored. For most of the next century, it was thought that space was empty and that a vacuum could be created simply by removing all matter. Then, late in the 1900s, it became apparent that a "vacuum" did contain a form of thermal radiation. Furthermore, even after the vacuum was cooled to absolute zero, there remained some form of nonthermal radiation.[3] It became clear that

Tesla's ideas had to be revisited. Most physicists today have come to realize that "empty space" is in fact a virtual ocean of energy. This energy flows all around and through us, since we ourselves, by virtue of the nature of atoms and molecules are mostly empty space.

Scalar, zero-point, and dark energy

Today, Tesla's "radiant energy" is referred to as "scalar energy," the term most frequently used by those in the natural sciences. However, it is also known as "zero-point energy" and sometimes "dark energy." The term "zero-point energy" was coined when scientists first determined that vacuums contain some form of nonthermal radiation.[3] This term is still used today by scientists studying the extraction of energy from the vacuum of space. The term "dark energy" was used very early in the development of electromagnetic theory to describe energy coming from the terminals of a battery that could not be accounted for in terms of electrical current. This term is still used by astrophysicists to describe the energy of outer space. Even though the terms used by various branches of science are different, they all refer to the energy that exists everywhere in the universe.

Although many scientific circles now accept the existence of scalar energy, it is still not completely understood. The literature states that scalar energy is created where two equal and opposite energy waves meet. Although standard measuring techniques indicate that these two energies will cancel each other out, the canceled frequencies leave a standing or stationary energy potential.[4] Less the energy is focused in some way, it will spread out benignly in all directions. One reason that scalar energy has not yet made its way into textbooks is that it is a very fine form of energy, and current scientific instruments are incapable of measuring it. However, its myriad effects attest to its existence.[5]

Valerie Hunt, the UCLA neurophysiologist who gave validity to the existence of the human energy field, has referred to scalar energy as the "wave of creation."[6(p313)] Her years of work with the human energy field and with healers have given her many opportunities to witness the powerful potential of scalar energy. In her latest book, she comments on one healer's work as follows:

> *When she first worked with a paralyzed person there were no spatial movements in the involved area. Soon [we] noted small shimmering movements in very local tissue, indicated by moving dimples that expanded to take in the paralyzed area. The skin changed to a pink color. To me this looked like cell regrowth before whole muscles were regenerated. Immediately, I knew this was scalar energy from the outside environment. About ten years ago I first demonstrated this to be a biological phenomenon and I showed how this energy facilitates tissue healing.*[6(p xviii)]

James Oschman discussed scalar energy and its applications in bodywork in his widely acclaimed book, *Energy Medicine: The Scientific Basis*.[7(p206–208)]

The dipole key

According to Tom Bearden, considered to be one of the world's leading authorities on scalar energy and the man who advised President Ronald Reagan on the US "Star Wars" technology (the Strategic Defense Initiative of 1983), current electromagnetic theory is based on a number of misconceptions. The most obvious of these is the assumption that the electrostatic potential of space is zero.[4]

Another misconception in electromagnetic theory is that batteries and generators provide the power for circuits. Bearden has explained that what batteries and generators really do is to separate internal charges to form a dipole, with positive charges being moved to one terminal and negative charges to the other. Once a dipole has been formed, scalar energy is organized from the surrounding energetic ocean. A tiny amount of that energy is picked up and directed by the polarized electrical circuit; the rest remains unused.[8]

Bearden also noted that the first scientists to quantify electromagnetic fields (Heaviside, Poynting, and Lorentz in the 1880s and early 1900s) referred to but could not explain the enormous quantity of energy pouring from the terminals of any battery or generator. The equations that they eventually developed took into consideration only the tiny component of energy flow that entered the circuit. In the end, they reasoned that the huge flow missing the circuit "had no physical significance." This is like saying that the wind blowing over the ocean has no physical significance, except for that small portion striking the sail of one sailboat. Nonetheless, those who study electrodynamics today continue to use those original equations, often with an incomplete understanding of their development. Bearden's paper, "Giant Negentropy from the Common Dipole," is an excellent, fully referenced review of the history of electromagnetic theory.[8]

With knowledge of the background and a review of the historical development of electromagnetic theory, it is easier to understand why the dipole is the key. It is, in fact, the gateway to an infinite amount of energy that, as Tesla predicted, could be accessible from any point in the universe. According to Bearden:

> *Any dipole has a scalar potential between its ends, as is well-known . . . The scalar potential decomposes into—and identically is—a harmonic set of bidirectional longitudinal electromagnetic wavepairs. Hence the formation of the dipole actually initiates the ongoing production of a harmonic set of such biwaves.*[9]
>
> *Once the source dipole is formed, . . . it induces the spreading giant negentropic reordering of the vacuum energy, extracts (transduces) EM [electromagnetic] energy from the continuously reordered vacuum, and pours out from the terminals of the generator (or battery) a vast 3-flow of EM field energy along the external circuit.*[8]

Thus, the separation of charges, creating two equal and opposite waveforms, organizes the vast sea of energy in space and makes electromagnetic energy available for use. Providing a pathway (a wire) for the organized energy literally changes and channels a small portion of the energy, allowing it to power electrical equipment.

The human dipole

It was Harold Burr, the Yale University neuroanatomist and editor of the *Yale Journal of Biology and Medicine*, who first noted the dipolar nature of the energy field around animals, plants, and humans. He claimed to be able to predict many things about a person's physical and emotional health by checking the voltage between head and hand.[10(p83,98)] As he described it, the entire human body is a dipole, with the positive pole at the head and the negative pole at the base of the torso (Figure 4A). This dipole is the source of the human bioenergy field, which is composed of scalar energy, organized and held in coherence by the polarity of the dipole.

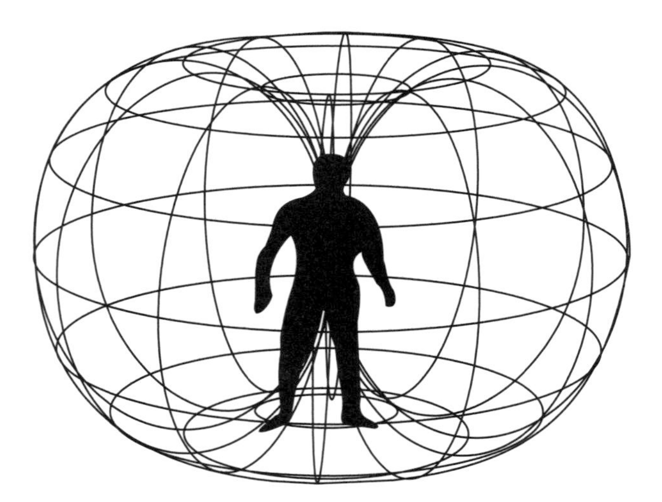

Figure 4A *Representation of the entire human body as a dipole*

Besides *being* a dipole itself, the human body is made up of billions of component dipoles. The scalar potential created by each tiny dipole provides electromagnetic energy for numerous systems and for the transmission of signals and information throughout the body. This explains how every cell can instantly know the activities of every other cell in the body. The medium through which information is transmitted is coherent scalar energy, supported at the physical level by the liquid crystalline continuum of the body.

Dipoles, and thus scalar energy, play a critical role in every process in the body.[11] Robert Becker, an orthopedic surgeon and coauthor of the book *The Body Electric*,[10] identified the role of dipoles in regeneration and wound-healing. He verified the "current of injury" and demonstrated changes and shifts in electrical polarity throughout the healing process.[10(p73)] His work indicated that both an appropriate dipole and an accompanying current are necessary for healing to occur.

The quartz connection

Bruce Lipton, PhD, professor of cell biology at the University of Wisconsin Medical School and author of *The Biology of Belief*,[12] determined that the cell membrane (a dipole) was the structural and functional equivalent of a silicon (quartz crystal) computer chip. His "eureka moment" was accompanied by the realization that, just like computers, the cell was programmed from outside itself, not by the DNA within.[12(p90–92)] In other words, cells receive their programming from the environment, from and through the energy field outside the body. If the information in that field is distorted, the cell receives distorted or diseased programming.

As determined by Hunt, healers and bodyworkers deliver coherent scalar energy, resonating at or near the frequency of the Earth (the Schumann resonance, 7.83 Hz), to areas of the bioenergy field that have become distorted because of trauma, whether physical or emotional.[6(p xviii),13(p66–67)] In a 2007 interview, Hunt said:

> *We have found that all medical conditions are improved or eliminated by bioscalar activation, although in some situations*

> *the results can take weeks or months to occur. But the direction is always positive as the body establishes new energy field patterns that are self-healing.*[14]

According to Hunt, coherent scalar energy delivered from outside the human body activates the bioscalar energy field and is capable of initiating and supporting the healing response. Its effects are positive and repeatable. The same scalar energy is emitted during other forms of energy work, including qigong.[7(p206)]

Nearly everyone is a healer to some degree, with the capacity to direct coherent scalar transmissions for healing. For example, every mother has intuitively placed her hands on her child's "boo-boo" for a soothing effect that we have come to discover is very real. For some, the ability to direct this energy is more pronounced, and almost anyone can develop the skill. CAMS devices direct coherent scalar energy in much the same way. They can make up for the difference between the unskilled healer in all of us and the most accomplished and recognized healers.

As described in the previous chapter, quartz crystals are transducers, capable of changing one form of energy into another. Their activation creates a dipole capable of extracting scalar energy from space and converting it, in small amounts, to electrical current. However, they can also function in the other direction, converting electromagnetic energy into scalar waves for use in the healing process. This is how crystals are used in the CAMS technology. The quartz crystals in a CAMS device convert electricity or light into coherent scalar energy that resonates at the Schumann resonance. This energy is directed and focused by the power of intention to specific areas of anticoherence in the human energy field. The next chapter discusses the power of intention.

References

1. Bearden T. Tesla's secret and the Soviet Tesla weapons. Self-published; 1981. Available from: *http://www.cheniere.org/books/part1/starting%20 pages.htm*

2. Valone T. Zero-point energy from the quantum vacuum [lecture on video]. Integrity Research Institute; 2004. Available from: *http://www.integrityresearchinstitute.org/ZPENERGY.html*. Quoting N. Tesla in a lecture delivered before the American Institute of Electrical Engineers and reprinted in *The Electrical World*, July 11, 1891.
3. Valone T. Inside zero point. Available from: *http://www.seaspower.com/InsideZeroPoint.htm*
4. Bearden T. The final secret of free energy. Self-published; 1993. Available from: *http://www.cheniere.org/techpapers/Final%20Secret%209%20Feb%201993/indexold.html*
5. Valone T. Zero-point energy from the quantum vacuum [lecture on video]. Integrity Research Institute; 2004. Available from: *http://www.integrityresearchinstitute.org/ZPENERGY.html*
6. Conrad E, Hunt V. *Life on land: the story of continuum*. North Atlantic Books; 2007.
7. Oschman J. *Energy medicine: the scientific basis*. Churchill Livingstone; 2000.
8. Bearden T. Giant negentropy from the common dipole. Self-published; 2000 June 6. Available from: *http://www.cheniere.org/techpapers/GiantNegentropy.pdf*
9. Bearden T. The unnecessary energy crisis: how to solve it quickly. Self-published; 2000 June 12. Available from: *http://www.cheniere.org/techpapers/*
10. Becker R, Selden G. *The body electric*. William Morrow and Co.; 1985.
11. Szent-Györgyi A. *Introduction to a submolecular biology*. Academic Press; 1960.
12. Lipton B. *The biology of belief: unleashing the power of consciousness, matter and miracles*. Mountain of Love/Elite Books; 2005.
13. Hunt V. *Infinite mind: science of the human vibrations of consciousness*. Malibu Publishing Co.; 1996.
14. Trivieri L Jr. The human energy field—an interview with Valerie Hunt, Ph.D. Part 4. *Health Plus Lett*. 2007;5(22). Available from: *http://www.1healthyworld.com/ezine/vol5no22.cfm*

Chapter 5

Intention

According to Deepak Chopra, a medical doctor and widely read author of many books on the mind–body connection, everything that happens in the universe begins with intention. In his January 2010 online newsletter, he wrote, "Your focused intentions set the infinite organizing power of the universe in motion."[1] Chopra's statement reflects the findings of the many scientists who have documented the power of intention and its potential for mobilizing the scalar energy of the universe.

William Tiller, professor emeritus of materials science and engineering at Stanford University, has spent many years studying the effects of intention. During a four-year period of research, he and his fellow researchers quantified the effects of intention-imprinted devices on experimental outcomes. Their book, *Conscious Acts of Creation: The Emergence of a New Physics*, describes that research. Ultimately, they concluded that "Focused intention can be made to act as a true thermodynamic potential and strongly influence experimental measurements."[2(p xi)]

Tiller's experiments demonstrated statistically significant and repeatable changes in the pH of water. He also recorded substantial changes in enzyme and coenzyme activity. Further experiments showed that intention produced marked changes in the larval development time of fruit flies.[2(p54–88,102–125,147–153)] Perhaps even more interesting was the effect that intention-imprinted devices had on the experimental space itself, the location where experiments were conducted. The longer the devices were operated in a particular area, the stronger the experimental effects became. In other words, the energy of the space itself was transformed or "conditioned" through ongoing intention.[2(p169–231)] In these experiments, conditioned spaces were shown to alter experimental outcomes for up to one year after the intention-imprinted devices had been removed. This lends credence to our developing understanding that the energy of space can be organized, and that when it is organized, it is capable of holding information and energy patterns.

Much research on heart-focused intention has been conducted at the Institute of HeartMath in Boulder Creek, California. One controlled, blinded study showed that heart-focused intention could wind or unwind the two strands of the DNA helix, in some cases causing a 25% increase in winding or unwinding relative to control samples.[3] This research also tested the effects of intention at a distance, whereby the intentions were sent from a location a half mile away. Changes were noted even at this distance, which indicates that intention is not confined by space.[3]

Glen Rein, PhD, director of Quantum Biology Research in Deer Park, New York, has also done considerable research on the power of intention to alter biological parameters. His studies showed that DNA synthesis in tumor cells could be altered by focused intention,[4] that intention could alter the molecular structure of water,[5] and that the information content associated with scalar energy could be stored in geometric patterns.[6] These findings support many theories about nature's use of geometry and patterns for creation. It also suggests the importance of the organized scalar field for transferring intentions at infinite speed, to any location.

Nonlocality

Valerie Hunt, the UCLA neurophysiologist whose work helped to define the human electromagnetic field, determined that focused intention was responsible for effects at a distance. Speaking of the ability to generate a healing response from a distance (remote healing), Hunt said, "One's energy can be controlled by thought. If one successfully projects energy via a thought-field, one can transmit energy over great distances with specific intent."[7(p269)]

The eminent Herbert Fröhlich (1905–1991), one of the great pioneers of superstate physics, who described the characteristics of the fully ordered state, found that once molecules reach a certain level of coherence, they take on quantum mechanical qualities, including nonlocality.[8(p49),9] The term "nonlocality" refers to the capacity to initiate something at one location that has measurable effects in another location, without any direct physical contact. Fröhlich described a system of molecular oscillators reaching a certain threshold frequency, at which point the whole system condensed to create one giant dipole mode. According to Fröhlich, this kind of order could accomplish the "lossless transmission of energy" that Tesla originally predicted[9] (see Chapter 4). This is similar to the organization within a laser, where coherence enhances energetic potential and creates a whole new set of possibilities. From Fröhlich's point of view, the fully ordered state accounts for nonlocal effects.

Coherence, described in Chapter 1 as a fixed-phase relationship between waves or particles, links together the particles, waves, and energetic potential in any system. When one wave or particle moves, all of the components respond at the same time with the same action, like a school of fish swimming in formation. Fröhlich's work demonstrated that, at a certain threshold, coherence was established at the quantum (subatomic) level, which in turn created the potential for instantaneous transmission of energy and which also explained nonlocal effects. This has since been referred to as "quantum coherence."[10] Mae-Wan Ho, author of the book *The Rainbow and the Worm: The Physics of Organisms* and director of the prestigious Institute of Science in Society, in London, England, provided this further definition of quantum coherence:

> *'Coherence' is generally understood as 'wholeness', a correlation over space and time. Atoms vibrating in phase, teams rowing in*

synchrony in a boat race, choirs singing harmony, troops dancing in exquisite formations, all conform to our ordinary notion of coherence.

Quantum coherence implies all that and more. Think of a gathering of consummate musicians playing jazz together ('quantum jazz') where every single player is freely improvising from moment to moment and yet keeping in tune and in rhythm with the spontaneity of the whole. It is a special kind of wholeness that maximizes both local freedom and global cohesion.[10]

Fröhlich recognized another important component necessary for nonlocal effects to take place: the presence of a dipole. A dipole not only organizes the energy, it also establishes direction (see Chapter 4). Focused intention creates a dipole and a pathway (bidirectional longitudinal waves) between the intention and the object of intention. In an instant, focused intention can be transmitted to any location, no matter the distance.

The concept of "entanglement" was introduced in 1935 by Austrian physicist Erwin Schrödinger to describe the phenomenon of nonlocality.[11] According to Stanford University's online encyclopedia of philosophy:

Quantum entanglement is a physical resource, like energy, associated with the peculiar nonclassical correlations that are possible between separated quantum systems. Entanglement can be measured, transformed, and purified. A pair of quantum systems in an entangled state can be used as a quantum information channel.[12]

Over the years, many other scientists have contributed to our understanding of entanglement, which explains nonlocal effects.[12] Ho, the researcher whose work confirmed the liquid crystallinity of living organisms, referred to entanglement as a wonderful way to describe the inseparable oneness that exists in the universe.[13]

Focus

The interesting thing about the research mentioned at the beginning of this chapter is that the intentions in each successful experiment were initiated by those with proven skills: healers, mediators, and those trained in heart-focused intention. Whereas in the HeartMath study, individuals with no previous training were unable to produce any change in the winding and unwinding of DNA, both Tiller and Rein utilized trained practitioners in their work.

The human mind is very powerful. However, certain skills may lie dormant until they are accessed and honed. The ability to focus and to manifest a healing outcome has, until now, been limited to a few individuals who either have exceptional gifts or have devoted the time and effort to the development of these skills. CAMS devices make this method of healing more widely available, as they are able to amplify and focus anyone's healing intention. The CAMS technology effectively makes up the difference between those who are skilled and those who simply believe in and participate in the process.

Believing in and participating in the process are acts that connect the user's thoughtful intent to heal with the one being treated. Pulsing the breath, as Marcel Vogel discovered and as others have confirmed,[14(p139)] gives the intention maximal focus. A pulsed breath consists of a quick release (forced exhalation) of the breath through the nose, like a mild snort. It is believed that the pulsed breath, a technique that originated in antiquity, coordinates the right and left hemispheres of the brain and thus presents a unified energetic output.

The CAMS device thus receives a 7.83-Hz electromagnetic signal. It amplifies and transduces the energy, producing a coherent scalar output that follows the direction generated by intent. The focused, amplified, coherent flow of energy resonating at the Schumann resonance works like a tuning fork to re-establish coherence in anticoherent areas of the human or other bioenergy fields.

Blocking the work

Although the human mind is very powerful, and although focused intention can direct powerful healing energy, the mind also has the capacity to block or minimize healing.

In her work with healers, Hunt noted that healing depended on the transaction between two people. According to Hunt, at least two factors are at play. First, the healer has to believe that he or she can accomplish the healing. Second, the person to be healed has to (at some level) want the assistance of the healer. If either of these two conditions is not satisfied, health cannot be restored.[7(p269)] In our own work with the CAMS devices, my colleagues and I have reached the same conclusion: the effect of the CAMS devices can be limited or nullified if one or both of these conditions are not met.

The practitioner's belief is the first factor. There is much to be said for the old axiom, "If you believe you can, you can; and if you believe you can't, you can't." Beliefs are the filters through which we see life. They touch every thought we think and every action we take. For example, if the person operating the CAMS device does not believe it will work (perhaps thinking that the condition being treated is incurable or that the problem is too advanced or perhaps even, at some level, not believing that the technology works at all), then they will be right.

Bruce Lipton, PhD, professor of cell biology at the University of Wisconsin Medical School, has shown, from a biological perspective, how beliefs influence everything we do and everything we think.[15] When operating a CAMS device, doubt or lack of belief creates a mixed message. The practitioner's conscious intention may be to send healing energy, but if there is disbelief at some level, his or her subconscious mind will override the conscious intention. Mixed messages create a disruption in the information delivered through the CAMS device, and the healing energy is nullified.

For those times when a practitioner may be addressing circumstances that are unfamiliar or perceived as being beyond his or her ability to help, a mind-set of *wonder* may be useful. In other words, the mind can be trained to think something like, "I wonder how this patient will make use of this healing energy?" or "I wonder how long it will take before this patient notices relief?" These kinds of thoughts open the door for possibility and allow for the clear transmission of positive healing intention from the practitioner.

For those individuals who may still be caught believing that the technology does not work or that they cannot make it work, reading this book will help. Working with the CAMS devices and having positive experiences, both personally and with patients, will also help in overcoming limiting beliefs and in establishing empowering beliefs related to use of the technology.

The second factor in the transaction between practitioner and patient resides with the patient, the one requesting assistance. If, at some level, the patient believes that he or she cannot heal, or if he or she is not ready to heal (for example, if there is secondary gain or a feeling of unworthiness), the receipt of coherent energy will be blocked. In fact, it is possible that a belief in disease or a belief in disease as punishment is the underlying factor in the first place. The patient may have been creating anticoherence in his or her energy field for a long time through underlying beliefs of this nature. In this case, the beliefs are themselves the cause of failure to heal.

Negative or limiting beliefs, just like physical trauma, can create anticoherence in the energy field. These disturbances, referred to as "interference fields," eventually cause physical or emotional problems. The good news is that locating and treating the appropriate interference field with one of the CAMS devices will solve many health problems caused by disempowering beliefs or emotions, for practitioners and for patients alike.

In the next chapter, we will explore the nature and significance of interference fields.

References

1. Chopra D. Harness the power of intention. *Agni Newsl.* 2010 Jan. Available from: *http://www.chopra.com/agni/Jan10/deepak*
2. Tiller W, Dibble W, Kohane M. *Conscious acts of creation: the emergence of a new physics.* Pavior Publishing; 2001.
3. McCraty R, Atkinson M, Tomasino D. Modulation of DNA conformation by heart-focused intention. Boulder Creek (CA): Institute of HeartMath; 2003. Publ No. 03-008. Available from: *http://www.heartmath.org/templates/ihm/section_includes/research/research-intuition/Modulation_of_DNA.pdf*
4. Rein G. Water memory: carrier of conscious intension [lecture on video, from 27th annual meeting]. Society for Scientific Exploration; 2008. Available from: *http://www.scientificexploration.org/talks/27th_annual/27th_annual_rein_water_memory_intention.html*
5. Rein G, et al. Structural changes in water and DNA associated with new physiologically measurable states. *J Sci Explor.* 1994;8(3):438–439.
6. Rein G. A bioassay for negative gaussian fields associated with geometric patterns. *Proc Acad New Energy.* 1997 May.
7. Hunt V. *Infinite mind: science of the human vibrations of consciousness.* Malibu Publishing Co.; 1996.
8. McTaggart L. *The field: the quest for the secret force of the universe.* Harper-Collins; 2002.
9. Fröhlich H. Evidence for Bose condensation-like excitation of coherent modes in biological systems. *Phys Lett.* 1975;51A:21.
10. Ho MW. Quantum phases and quantum coherence. ISIS Rep. No. 17/03/04. London (UK): Institute of Science in Society. Available from: *http://www.i-sis.org.uk/QPAQC.php*
11. Schrödinger E. Die gegenwärtige Situation in der Quantenmechanik [The present situation in quantum mechanics]. *Naturwissenschaften.* 1935 Nov.
12. Bub J. Quantum entanglement and information. In: Zalta EN, editor. *The Stanford encyclopedia of philosophy.* Summer 2009 ed. Available from: *http://plato.stanford.edu/archives/sum2009/entries/qt-entangle*

MW. The organic revolution in science [lecture]. Bioneers Conence; 1999 Oct 29–31; San Francisco (CA). Institute of Science in Society. Available from: *http://www.ratical.com/co-globalize/MaeWanHo/organic.html*

14. Comeaux Z. *Robert Fulford, D.O. and the philosopher physician.* Eastland Press; 2002.

15. Lipton B. *The biology of belief: unleashing the power of consciousness, matter and miracles.* Mountain of Love/Elite Books; 2005.

Chapter 6

Interference Fields

Interference fields are disturbances or areas of anticoherence in the body's bioenergy field. They are caused by trauma, either physical or emotional. They can even result from limiting belief patterns, which usually develop early in life during moments of trauma or stress and which are held in the body's energy field. Interference fields eventually affect the physical body, manifesting in a wide variety of symptoms, such as pain, inflammation, fatigue, anxiety, digestive problems, depression, and many diseases and other conditions.[1(p24–32)]

The discovery of interference fields is credited to two German physician–dentists Ferdinand and Walter Huneke, who were active in the 1930s and 1940s. Years of investigation, which began with use of a local anesthetic to treat their sister's chronic migraine headaches, resulted in the discovery that pain sometimes disappeared immediately, through a phenomenon referred to as a "flash phenomenon" or "lightning reaction," discussed later in this chapter, when an appropriate interference field was treated.[2(p84–85)] They also determined that pain in one part of the body could result from an interference field in a distant area of the body. Unfortunately, all of the work of the Huneke brothers was published in German, and only recently has some of it been translated into English. As a result, few doctors outside Europe are aware of its significance.

The presence of interference fields was implied by Harold Burr in the 1940s and noted by Valerie Hunt in the 1970s (see Chapter 1).

Both of these researchers recognized the role of interference fields as the underlying cause of illness. Burr, Hunt, and the Huneke brothers were also aware of the electrical nature of the disturbances caused by interference fields.

Interference fields as electrical disturbances

The Huneke brothers discovered that the cell membranes in tissues affected by interference fields have an abnormal resting potential.[2(p84–85)] At rest, the normal cell membrane is polarized, maintaining a negative interior charge and a positive exterior charge, similar to the electrodes of a battery. In other words, the cell is a dipole. Cells within an interference field generally have lower electrical potentials than the cells of surrounding tissues.[1,2(p27–34)] This means that these tissues are less able to respond to external stimuli and are also less able to react to the metabolic information transmitted via the autonomic nervous system. In essence, the tissues influenced by an interference field become cut off from their surrounding tissues.[2(p32–34)] Once isolated, they are no longer nourished (energetically or nutritionally), and they can eventually become a source of inflammation, pain, or other pathology.[2(p27)]

Changes in the electrical potential of the cell membrane may also cause changes in its permeability. As a result, toxins accumulate within the cells, causing a feedback loop, whereby the cells are prevented from restoring their normal electrical potential.[2(p28)] Much research has been conducted to determine the consequences of changes in membrane permeability. One researcher explained it this way:

> *The more we look into problems with permeability, the more we get the confirmation that at the beginning of nearly every disease process we discover membrane damage or change in membrane permeability.*[3]

Thus, the significance of interference fields becomes obvious from an entirely new perspective. When allowed to progress to the physical level, interference fields cause electrical imbalances in membrane electrical potential, which result in changes in membrane permeability, both of which have been linked with pain, inflammation, and disease.

Neural therapy

The Huneke brothers developed neural therapy as a method to treat interference fields and to restore the normal resting potential of the cell membranes in the involved tissues. Despite its name, neural therapy is not necessarily a treatment of nerves. However, it does restore the functioning of the autonomic nervous system, which is always dysfunctional (in a state of dysautonomia) within the area of an interference field.[1(p24)]

The basis of neural therapy is the fact that local anesthetics, particularly those used by the dental profession, have an interesting ability to regulate unstable electrical membranes in living tissue. These local anesthetics temporarily hyperpolarize both healthy and abnormal cell membranes. When the effects of the anesthetic wear off, any previously abnormal membrane resting potential returns to a level closer to normal, as illustrated in Figure 6A.[1(p39)]

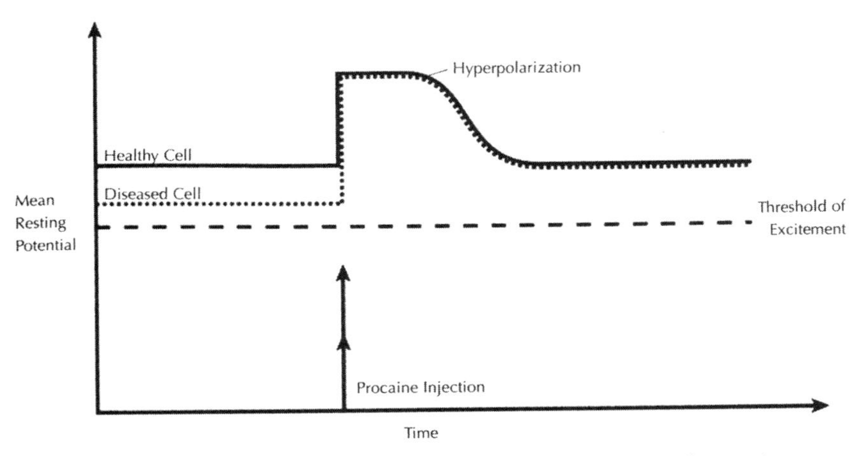

Figure 6A *Effect of procaine on resting potential of the cell membrane*

The local anesthetic procaine hyperpolarizes both healthy and abnormal cell membranes. After the effect wears off, the membrane potential returns to a level closer to normal.

(Reproduced from *Neural Therapy*, by Robert F. Kidd—used with permission.)

Neural therapy consists of injecting a local anesthetic into the tissues of an interference field. The treatment restores normal cell membrane electrical potential and results in stabilization of the tissues. A reduction of inflammation, pain, and other symptoms ensues.

In the recently translated 1995 edition of *Manual of Neural Therapy According to Huneke*, Peter Dosch explained that the use of local anesthetics allows cells to recharge and reset themselves under their own power. Then, with the disappearance of pain, the reactive inflammation also disappears.[2(p28)]

Treating interference fields using the techniques developed by the Huneke brothers and advanced by other physicians in Europe is common practice in Germany and other parts of Europe. This is evidenced by the fact that *Manual of Neural Therapy According to Huneke*, the most widely read German book on the subject, is now in its 14th edition.[2] Robert Kidd, an independent practitioner working in Ontario, Canada, has done much to bring awareness of this profound therapy to the English-speaking world. His book, *Neural Therapy: Applied Neurophysiology and Other Topics*,[1] provides detailed information on interference fields and their treatment.

Kidd has found that the CAMS technology is as effective as local anesthetics for the treatment of interference fields. In many ways, the CAMS technology is superior because it does not require injections. In fact, no physical contact is necessary. In Kidd's words:

> *A more recent invention is the Tenscam device . . . The device is held approximately 18 inches away from the body and is pointed at the interference field. The pulse generator is turned on, and the interference field is stimulated for 1 or 2 minutes. Most patients feel nothing, but a few sense a barely detectable tingling or heat. The therapeutic response is immediate and as clear-cut as that achieved with procaine injections. The response can be confirmed by autonomic response testing or (as appropriate) by alterations in skin temperature or by changes in ultrasonographic findings.*[1(p65)]

Kidd's book includes photographs of the immediate response achieved using a CAMS device. Since his discovery of this technology, Kidd has found it rarely necessary or desirable to use local anesthetics in his neural therapy practice (R. Kidd, personal communication).

Development of interference fields

Interestingly, one of the best-known locations of interference fields is surgical scars,[4(p164–165)] especially scars resulting from surgical procedures performed during periods of emotional distress.[1(p35)] Of course, not all scars result in interference fields, but work outlined in Alfred Pischinger's book, *The Extracellular Matrix and Ground Regulation*, demonstrated that a high percentage of scars exhibit increased resistance (abnormal membrane electrical potential),[4(p165)] with the possibility of causing physical symptoms over time. Kidd explained that although interference fields involve some sort of trauma or injury, they are also typically accompanied by a "complicating factor," such as a delay in healing or a secondary infection.[1(p26)] As noted above, a common complicating factor is emotional distress at the time of, or directly after, the injury or trauma. Scars or portions of scars that are still tender or that remain unusually red after healing are likely to harbor interference fields.

Scars are not always visible on the surface of the body. Internal scars may result from surgery or may be caused by chronic inflammation, which occurs with bowel diseases, arthritis, and infection. When the internal scarring becomes significant, a "neurological tipping point" is reached. The disturbed energy fields of these internal scars will cut off entire areas of the body from the regulation of the autonomic nervous system.

Interference fields are also often found in the teeth (as a result of cavities, dental work, extractions, or infection), the tonsils, and the autonomic ganglia (nerve "junction boxes"), which serve stressed organs of the body such as the liver or the digestive organs. Interference fields may also be found in the stressed organs themselves, at sites of mechanical joint or muscle stress, at sites of nerve entrapment, and even at puncture sites.[1(p25–32)] The practitioner should consider each of these possibilities when attempting to locate the interference fields associated with distress.

As previously noted, many if not most physical symptoms have an emotional component. In cases with any emotional involvement, regardless of the circumstances that caused its development, one particular interference field is nearly always present, located on top of the head, above and behind the right ear, as shown in Figure 6B. Treatment of this interference field can resolve many seemingly unrelated difficulties in areas throughout the body. For novice and advanced practitioners alike, this interference field is a good place to begin treatment.

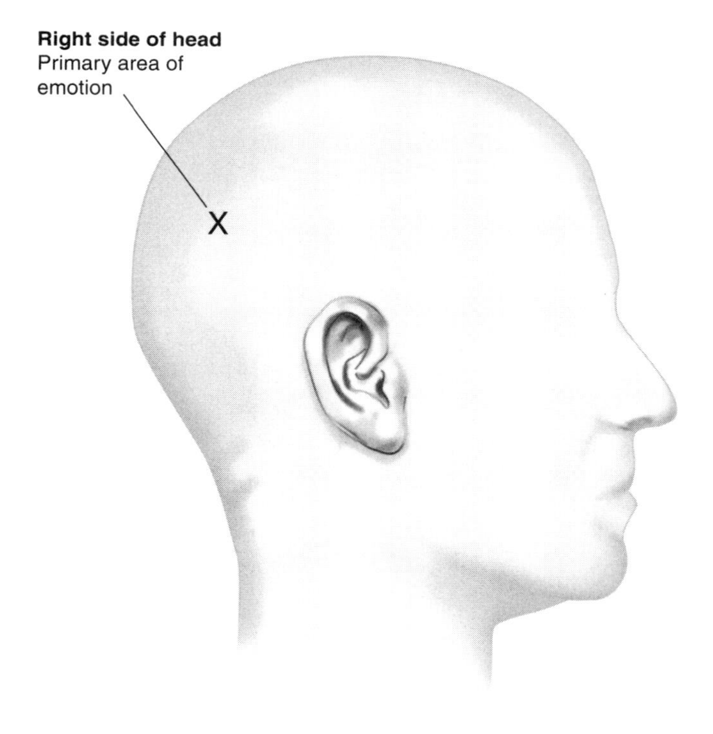

Figure 6B *Location of the interference field indicative of unresolved emotional conflicts*

Response time and lightning reactions

The effects of treating an interference field are immediate. They can be confirmed by changes in skin temperature and color, by autonomic response testing (discussed in Chapter 7), by galvanometric testing, and by ultrasonographic testing.[1(p65,67)] Each of these methods verifies significant and immediate changes in the cellular environment. Occasionally, *relief* from symptoms is also immediate. This was referred to by the Huneke brothers as a "flash phenomenon" or "lightning reaction." These reactions are more likely to occur with treatment of recent injuries or treatment of otherwise healthy individuals.

The CAMS devices are ideal for treatment of recent injury or other trauma. Under these circumstances, lightning reactions are almost to be expected, because the affected tissues have not been in a state of dysfunction for very long. Restoration of coherence in the energy field re-establishes normal membrane electrical potential, which reduces inflammation and pain. Complications are minimal, and many patients can resume their normal activities immediately. The implications are obvious for sports injuries. Jeffrey Spencer, a member of the US Cycling Federation Elite Medical Staff and chiropractor for Lance Armstrong, is a proponent of the CAMS technology, and a CAMS device was used during the last two years of Armstrong's record-setting seven wins of the Tour de France (J. Spencer, personal communication; photographs on file with author).

If an interference field has existed for a long time, or if the disturbance in the bioenergy field has caused long-term chronic symptoms, a single treatment may bring only minimal, temporary relief. However, *any* relief is a sign that an appropriate interference field has been treated. Subsequent treatments typically produce longer responses, but every individual is different. Regular treatment of long-standing interference fields (daily or several times a week) may be necessary to retrain tissues that have been trapped in dysfunction for a long time. Some conditions require significant retraining, and months of daily or weekly treatment may be involved. Under these circumstances, it is helpful if the patient has access to a handheld Personal Tuner or other CAMS device.

More often than not, a noticeable response to treatment occurs within the first 24 hours. Even a very short response is encouraging and indicates that treatments should be continued. The absence of a noticeable response may indicate that the treated interference field was not a significant contributor to the symptoms and that another interference field may be connected to the presenting problem. However, lack of a noticeable response can also mean a variety of other things, depending on the individual and the symptoms being treated. For example, some conditions have symptoms that are not easily noticed, especially if there is no overt pain. In such cases, several treatments may be necessary before a noticeable response takes place. Patients should be instructed to pay attention to changes in their level of energy, mood, range of motion, and other potential indicators. Unlike most traditional medical interventions, a CAMS treatment can cause no harm, and there are no contraindications to repeat treatments (except in pregnancy).

Conditions that may prevent successful treatment

Sometimes, repeat treatments yield no response, even when all indications are that an appropriate interference field is being treated. The patient may be "blocking" the work through lack of belief in its efficacy, but there are a number of other reasons for a lack of response to a CAMS treatment. Most of these reasons are related to our modern lifestyle and the environment in which we live. Several authors, including Dosch and Kidd, have speculated that one of the reasons that lightning reactions seem less common today than they were in the Huneke brothers' day is because of changes in our food sources and in our environment.

Generally speaking, the quality of our food is much lower today than it was in the 1940s. Most of today's food is grown on mineral-deficient, chemically fertilized soils. It is laden with pesticides, preservatives, additives, and environmental poisons. It is often irradiated and processed to the point that it is nearly devoid of nutritive value. This puts much of the world's population in a state of subclinical malnutrition that hinders the body's innate ability to respond to restorative treatments. Vitamin and mineral deficiencies, as well as deficiencies in essential fatty acids, natural antioxidants, and other nutritional elements, may play a role in a patient's lack of response to CAMS treatment. Unless these nutritional deficiencies are corrected, interference fields will continue to recur, shortening the duration of response or nullifying the effects of the therapy.

Another factor that is influenced by our modern lifestyle and that may prevent the body from making adequate use of CAMS treatment is the presence of toxins. Toxins may take many forms. Drugs (prescribed or over-the-counter; also including tobacco and alcohol) represent toxins that are prevalent today. According to Kidd, "The most common reason for poor response to neural therapy treatment is the presence of a drug. Any drug with a prefix of 'anti'—such as antibiotic, anti-inflammatories, antidepressants or antihypertensives—tends to block the autonomic nervous system."[5] Individuals who have been taking medications for a long time typically have a slower response to treatment or they may not at all.

Heavy metals are also toxins. Patients' bodies may contain mercury in dental fillings, lead from environmental pollutants, and cadmium, aluminum, arsenic, and other heavy metals from a variety of sources. Mercury is a neurotoxin that can poison the autonomic nervous system and defeat CAMS treatments.

Many other environmental toxins are prevalent in the modern world, found in air, water, and food; in cleaning agents, paints, and building materials; in personal care products and clothing; and in plastic bottles and food containers. Some of these products off-gas volatile compounds that build up over time in the human body. They have been linked with a growing list of "conditions" and "syndromes" that were unheard of 50 years ago. Many of these environmental toxins can be passed on from one generation to the next, conveyed from mother to child through the umbilical cord. Environmental toxins can nullify or reduce responses to CAMS treatment, or cause the responses to decline.

Another commonly overlooked reason for lack of response to treatment is dehydration, which is an important factor in cell membrane stability.[6] Notably, dehydration is not always about having *enough* water. Just as often, the problem relates to an electrolyte imbalance (i.e., mineral deficiency). Electrolytes, both within and outside of the cells, maintain the electrical potential of the cell membrane. In the absence of the appropriate minerals, the volume of the water in the extracellular environment (outside the cell) must be reduced to maintain the correct balance. When minerals, especially sodium, are lacking, the body allows water to pass directly through the organs of elimination without hydration of the tissues. This process leads to reduced circulation in the tissues and hence to the symptoms of dehydration: cold hands and feet, dry skin, digestive problems, and fatigue. One easy way to help determine whether a patient is dehydrated is to pinch the skin on the back of the hand—it should flatten again within 1 or 2 seconds.[6]

It is not within the scope of this book to discuss nutrition or detoxification. Many good books and websites with detailed information on these subjects are available to help patients to address these issues. In addition, many practitioners are already trained in these areas. Suffice it to say that when CAMS treatments *appear to fail* or when positive initial responses decline, there may be an underlying issue that is affecting the electrical stability of the cell membrane. Kidd has referred to this phenomenon as "generalized cell membrane instability." It can affect nearly every cell in the body[7] and is caused by nutritional deficiency, toxicity, and/or dehydration. These issues must be addressed in some manner before CAMS treatments will be successful. That said, treatment with the CAMS technology can be supportive while these issues are being [illegible]d.

Distant effects

When a broken bone is left untreated, it may not heal in its original position. If the healed bone is allowed to remain in an abnormal position, other areas of the body will eventually have to compensate, limiting mobility, reducing strength, and causing pain and other possible complications. However, if the bone is set back to its original position before healing, full mobility and strength can return. The same is true of the human bioenergy field. When a disturbance in one location is allowed to persist for a long period, it may cause dysfunction and pain in other areas of the body.

A given interference field can cause symptoms almost anywhere in the body. Often, these symptoms are far from the place that is exhibiting pain or disease. This is especially true of chronic pain, where abnormal electrical signals can be transmitted, via the autonomic nervous system, from the area of disturbance to distant areas of the body.[1(p7–8)] For example, an old appendix scar can cause migraine headaches, extraction of a wisdom tooth can cause chronic low-back pain, and arthroscopic knee surgery can cause shoulder pain. Symptoms in distant parts of the body are known as "referred pain," a phenomenon that is well recognized in internal medicine. [8(p93)] Physicians involved in pain management are well aware that chronic pain is often referred from a distant part of the body. The classic example of referred pain is the pain in the left arm experienced by people with impending myocardial infarction (heart attack).

Ferdinand Huneke observed the effects of a distant interference field for the first time when treating a woman with arthritis in her right shoulder. The woman's condition had resisted many previous treatments by other doctors. Huneke treated the right shoulder joint, as well as several other locations, with injections of procaine—a treatment that had been effective in similar cases—to no avail. However, several weeks later, a severe infection developed in an old osteomyelitis scar on the woman's left shin. She returned to Huneke in great pain, and he treated the scar itself. Immediately, the pain in her right shoulder subsided, for the first time in many years. She regained full range of motion and experienced complete resolution of her symptoms. Huneke wrote:

> *This experience was so startling that I could have no doubt that I was looking at a fundamentally new piece of knowledge and that I was on the track of a hitherto unknown law in the field of focal processes.*[2(p47)]

Pain and/or dysfunction in a distant area of the body can also result from blockage of an acupuncture meridian, whereby a distant organ along the blocked meridian is affected.[1(p10,24)] Because these relationships are difficult to predict, it is important to thoroughly scan the patient's entire body for the presence of interference fields. The most prominent will be clearly evident and should be treated first.

The next chapter describes various ways of locating interference fields.

References

1. Kidd R. *Neural therapy: applied neurophysiology and other topics.* Custom Printers; 2005.
2. Dosch JP, Dosch M. *Manual of neural therapy according to Huneke.* 2nd English ed. (translation of 14th German ed.). Gutberlet R, translator. Thieme Medical Publishers; 2007.
3. Dosch JP, Dosch M. *Manual of neural therapy according to Huneke.* 2nd English ed. (translation of 14th German ed.). Gutberlet R, translator. Thieme Medical Publishers; 2007. p. 28. Quoting Eppinger L.
4. Pischinger A. *The extracellular matrix and ground regulation: basis for a holistic biological medicine.* Eibl I, translator. North Atlantic Books; 2007.
5. Kidd R. What is neural therapy? [Internet]. Self-published; 2007. Available from: *http://www.neuraltherapybook.com/learnabout.php*
6. Kidd R. Dehydration. *Neural Ther Pract* [electronic newsletter]. 2010 Jan;5(1). Available from: *http://www.neuraltherapybook.com/newsletters/5-1.php*

7. Kidd R. Generalized cell membrane instability. *Neural Ther Pract* [electronic newsletter]. 2007 Sep;2(9). Available from: *http://www.neuraltherapybook.com/newsletters/2-9.php*

8. Silen W. Abdominal pain. In: Fauci AS, Braunwald E, Kasper DL, et al., editors. *Harrison's principles of internal medicine.* 17th ed. McGraw-Hill Companies, Inc.; 2008. p. 91–94.

Chapter 7

How to Locate Interference Fields

Interference fields can be located in a number of ways. Many practitioners find it helpful to begin by obtaining from the patient a thorough history of his or her injuries, illnesses, dental procedures, surgeries, unresolved emotional conflicts, and other potentially relevant traumas, beginning at birth. Any injury, surgical procedure, or other event that occurred in the weeks or months before the onset of symptoms should be suspect. However, even small injuries and long-forgotten illnesses, including seemingly unimportant emotional upsets, especially those occurring during the same time frame as other traumas, can lead to the development of significant interference fields. Patients may leave out critical information during a review of their medical history because they have forgotten an incident or believe it to be irrelevant. Obtaining the patient's medical history by interview is valuable, but if it is the *only* method used, the practitioner will have limited success in locating interference fields.

Interference fields can also be detected by so-called energetic testing methods and equipment-based methods. Although energetic testing is the subject of some controversy, it is quick and reliable once a certain level of skill has been acquired. With a bit of practice, anyone can learn the techniques described here.

Energetic testing

The term "energetic testing" is an umbrella term for a variety of techniques used by many health practitioners to interpret signals delivered through the autonomic nervous system. These testing methods are based on muscle strength, arm or leg length, skin temperature, and the condition of the bioenergy field, all of which can be determined by the trained practitioner without expensive equipment. The information gathered through energetic testing helps to address the body's needs in the present moment, without having to wait for laboratory reports or other analyses. This means that the body can also be retested following treatment to determine whether therapy has been successful and/or complete. Two forms of energetic testing are particularly suitable for the detection of interference fields: Manual Energy Evaluation and Autonomic Response Testing.

Manual Energy Evaluation

Manual Energy Evaluation (MEE) is my own version of a technique originally developed by the French osteopathic physician Jean-Pierre Barral. Barral's technique (initially known as manual thermal diagnosis and more recently as manual thermal evaluation) focuses on the thermal (heat) differences between healthy and unhealthy tissue. Although MEE is an offshoot of Barral's technique, it is not limited to the detection of thermal variations. In addition to detecting variations in the body's thermal status, it also reveals variations in the surrounding bioenergy field. Changes in these parameters help in identifying areas of anticoherence (i.e., interference fields).

The human hand is a surprisingly good detector of energy. However, when it is used to physically touch an object, its ability to distinguish energetic changes is masked by tactile perception. When the hand is kept at a distance, tiny differences in temperature and other energetic parameters become apparent. For example, the hand is capable of distinguishing differences in temperature of less than 1/10 of 1°C between two objects.[1(p18)] It is therefore an effective "scanner" for energetic changes. The area in the center of the dominant hand, just below the wrist but not quite at the center of the palm, is the most sensitive part of the hand for most people (see Figure 7A). This area is used in the same way that a scanning device is used to scan the body of a patient.

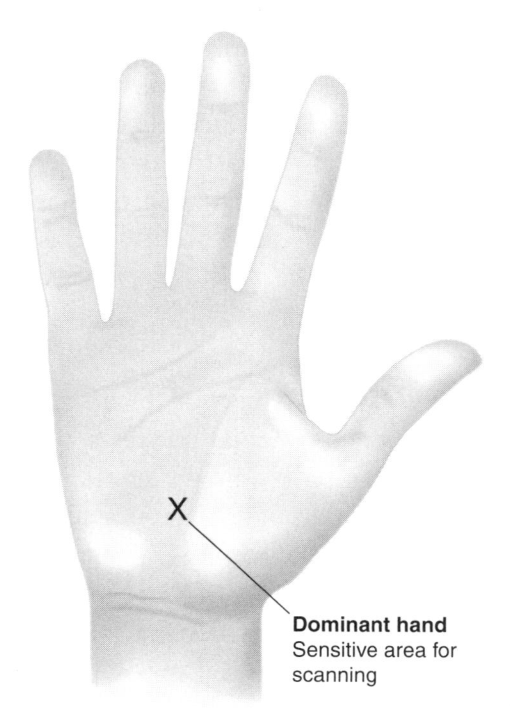

Figure 7A *Use of the hand to detect an energy field*

The palmar area of the dominant hand—just below the wrist—is the most sensitive part of the hand for most people.

The human body radiates infrared heat and scalar energy. Although clothing blocks the transfer of infrared radiation, the hand is also sensitive to wavelengths somewhat longer than those in the infrared part of the spectrum. This fact accounts for the hand's ability to scan the body through clothing that would be insulating to heat alone.

MEE scanning

Dysfunction and disease create changes in the emission of heat from the body, which may be either greater or less than normal. These changes may be due to inflammation, circulatory disturbances, variations in metabolic activity, or other factors, all of which are related to altered membrane electrical potential. It is the *change* that is significant in MEE. Comparing one side of the body to the other side is the best way to identify any thermal changes associated with interference fields. Barral found that if the scanning hand remained stationary over one area of the body for too long, heat built up between the hand and the body, and a false reading was obtained. He noted that scanning was most effective when the hand was moving at a rate of about 2 feet per second.

Barral's technique focuses on the detection of thermal changes and is performed at a distance of about 4 inches from the body. However, with MEE, the hand *must* be positioned 18 to 24 inches from the body. At this greater distance, the hand can perceive the primary changes in temperature and disturbances in the energy field.

The human energy field is said to have several layers. Where each of these layers ends and the next begins is irrelevant to the effectiveness of MEE. Distortions in the energy field penetrate all of its layers and are best felt at the outer edge, 18 to 24 inches from the physical body. Interestingly, this is what most people define as their "personal space."

Practitioners who are new to MEE sometimes tend to let the scanning hand get too close to the patient's body. Inside the 18- to 24-inch distance, the practitioner loses contact with the energetic "resiliency" (a kind of cushiony resistance) that can be felt at the outer edge of the energy field. When this happens, the practitioner can no longer feel the energetic distortions and may connect with only a part of the available information. Therefore, especially when first learning MEE, the practitioner should pay careful attention to keeping the hand outside the energy field, where he or she will eventually acquire sensitivity to its subtle variations. Distortions in the bioenergy field can be felt either as indentations (troughs) or as vectors (spikes) in the energy, either of which indicates the presence of an interference field. The other important factor to remember is the need to keep the hand moving.

With MEE, it is the *difference* in temperature between one side of the body and the other and the *difference* in the "feel" of the energy between one side of the body's field and the other that are clues to the interference fields. Therefore, during scanning, attention should be paid to differences in temperature and in the energy field itself. While learning to perform MEE scanning, it may be helpful to close the eyes or to look away, to minimize visual distractions. Each practitioner must develop his or her own "sense" and "feel" of the energy. As with most skills, practice is the key.

MEE is completely noninvasive. In fact, it is a requirement that there be *no* physical contact with the patient. In today's world, this aspect of scanning is highly desirable, for the patient, as well as the practitioner. Further details about MEE are provided on the DVD *TensCam + Thermal*, available from Crosby Advanced Medical Systems. Workshops are also held periodically in the United States and overseas. Visit www.tenscam.com for details.

Autonomic Response Testing

In 1964, George Goodheart, an astute American chiropractor, noticed that weakness in certain muscles was correlated with dysfunction in certain organs. He mapped these correlations and developed an energetic testing technique called Applied Kinesiology. This method uses the body's muscle tone as a feedback system to identify the location of dysfunction and to determine appropriate treatment options. Since originating this technique, Goodheart and many of his colleagues have developed numerous adaptations, including "Touch for Health," Behavioral Kinesiology, Neuro-Emotional Technique (also known as NET), and Clinical Kinesiology. In 1993, Dietrich Klinghardt, MD, PhD, and Louisa Williams, DC, codeveloped neural kinesiology, which is based on the principles of Applied Kinesiology as they relate to neural therapy, specifically the identification of interference fields. The outgrowth of that development eventually resulted in Autonomic Response Testing (ART), the widely used form of kinesiology employed by many neural therapists today.

The basic ART technique involves having the patient lie down in front of the practitioner with one arm raised to the vertical position. The deltoid muscle in the shoulder becomes the "indicator" muscle. Light pressure is applied to the arm by the practitioner, who also attempts to pull the arm down toward the body. The patient is asked to resist this pressure during testing. The practitioner places his or her other hand over organs, nerve ganglia, scars, and other potential interference fields, and the patient's arm muscle will remain strong until an offending interference field is found. At that point, the same amount of pressure on the arm will noticeably weaken the muscle, and the arm will lose its ability to resist.

This explanation of ART is highly simplified, and there is much more to the successful application of this technique than can be presented here. One of the concerns about using ART is that the production of reliable results depends on the autonomic nervous system being in a responsive state. If the body is dealing with an overload of toxins or infection, if the body is dehydrated, or if the patient is preoccupied with an emotional upset, the autonomic nervous system will be incapable of responding to the stimuli used in ART. In this situation, the readings may be inaccurate. ART includes ways of dealing with these problems so that accurate readings can be obtained, but training is required for complete understanding of these methods.

Practitioners who are already trained in one of the forms of kinesiology will find it easy to use ART to find interference fields. Those without such prior training are encouraged to attend one of many workshops and seminars available to learn the nuances of ART. Alternatively, it may be preferable to focus on learning MEE. In my personal experience, I have found that MEE takes less time to learn and is ultimately more effective than many other methods.

Other energetic testing methods

Many practitioners are trained in the use of other energetic testing methods that may be equally useful for detecting interference fields. Anyone with such training should use his or her preferred method. However, it should be kept in mind that many of these methods also require that the autonomic nervous system be in a responsive state, without hindrance by toxins, infections, emotions, dehydration, etc.

Equipment-based evaluation

Other methods can be used to actually *quantify* the changes in cell membrane electrical potential and/or the changes in thermal emissions associated with interference fields. These equipment-based methods are helpful for patients who are not comfortable with energetic testing methods. They are also good ways to validate the effectiveness of the CAMS devices for patients who have difficulty accepting energy medicine. In some cases, practitioners may already have access to these types of equipment, several of which are discussed in the following sections. These methods should be used as the practitioner deems appropriate.

Electro-acupuncture

In principle, it is possible to measure the electrical resistance of the skin with any standard voltmeter or ohmmeter. Resistance at the surface of the skin reflects disturbances in cellular electrical potential (see Chapter 6). In 1955, the German physician Reinhold Voll originated a diagnostic method that combined the principles of acupuncture with modern electronics. Voll found that the skin is generally very resistant to electrical current, but specific locations are more conductive. These places correspond to acupuncture points, and their properties make it possible to measure the functional connections between the acupuncture meridians, the organs, and the other tissues of the body. Skin resistance can also be used as an indirect method of measuring the "energetic system" of the body and it can be used to locate interference fields.[2(p118),3]

Voll's method, known as EAV (for "electro-acupuncture according to Voll"), combines an ohmmeter, a diagnostic probe, and an electrode (which is held by the patient) to form an electrical circuit. Normal, intact skin has a resistance ranging from 150 to 500 kilo-ohms (kΩ). The resistance of skin over interference fields is generally greater than that of the surrounding skin, with readings between 600 and 1500 kΩ. Rarely will the resistance be below 40 kΩ.[2(p118)]

Since Voll's initial development of the concept, there have been many variations on his work, and a variety of devices have been created. These variations are all grouped into a class of diagnostic equipment known as EAV devices or sometimes electro-dermal screening (EDS) devices. Any of these devices can be used by trained practitioners to detect interference fields.

Thermography

Thermography measures the tiny changes in skin temperature that are associated with inflammation, circulatory disturbances, and other metabolic changes found within interference fields. Thermographic devices range from handheld sensors that measure skin temperature from a distance of several inches to thermographic cameras that convert temperature readings into images showing the hot and cold spots on the surface of the body. Sometimes, a small handheld thermographic device is an easy way to generate evidence of the cellular changes that occur rapidly with treatment of interference fields. However, it should be kept in mind that although most interference fields exhibit measurable changes in skin temperature, some do not.

CAMS device

Interestingly, another equipment-based method that has been used successfully by a number of practitioners to detect interference fields is the CAMS device itself. In the nonflashing mode, it can be used as a feedback system to pick up anticoherence in the bioenergy field and to amplify and relay it to the hand of the person holding the device. An interference field is indicated by a slight buzz or vibration that is felt when pointing the CAMS device at an area of anticoherence, from a distance of 18 to 24 inches outside the physical body. As with any energetic testing method, some practitioners will be more sensitive to this subtle vibration than others.

As Dosch explained in the *Manual of Neural Therapy According to Huneke*, "No equipment-based method of examination can ever be better than the physician who evaluates the quantitative results it produces."[2(p118)] Dosch was referring here to the practitioner's knowledge and experience, as well as to his or her intuitive sense, which are integral parts of the work. Knowledge, experience, and intuition guide the practitioner in using the information that has been gathered. Here, these faculties help the practitioner to determine which interference field or fields are creating the problem under investigation and which should be treated first. The next chapter focuses on selection and treatment of interference fields and includes a discussion of the use of the CAMS devices in such treatment.

References

1. Barral JP, Mercier P. *Visceral manipulation*. Revised English ed. Eastland Press; 2006.
2. Dosch JP, Dosch M. *Manual of neural therapy according to Huneke*. 2nd English ed. (translation of 14th German ed.). Gutberlet R, translator. Thieme Medical Publishers; 2007.
3. Block EF 4th. The basis of clinical diagnosis in bioelectromagnetic medicine. *J Bioelectromagn Med.* 2006 Jan;13. Available from: *http://www.diamondhead.net/p13.htm*

Chapter 8

How to Treat Interference Fields with the CAMS Devices

Most people have one or more interference fields.[1(p423)] According to Louisa Williams, codeveloper of neural kinesiology (the precursor of Autonomic Response Testing, as discussed in Chapter 7), "It is quite common nowadays for patients to have foci [interference fields] in the double digits."[1(p423)] Estimates vary depending on the definition of interference field that is used. Some practitioners restrict their estimates of numbers to interference fields that are associated with active symptoms, whereas others include subclinical or dormant interference fields.

Because of the body's capacity to compensate for many disturbances and deficiencies, many interference fields are of a subclinical nature—without symptoms. It is only when these disturbances approach a critical level that symptoms begin to manifest. Subclinical interference fields can become clinical (showing active symptoms) at any time, but such a shift usually follows some kind of additional stress. A person who has a subclinical interference field for many years may suddenly, for a variety of reasons (illness, a new job, loss of a family member, pregnancy, accident, etc.), experience symptoms.

Weston A. Price, a dentist who did a great deal of work to substantiate the connection between interference fields in the teeth and symptoms at distant locations, observed that under stress, a dormant root-canal tooth (a tooth that has previously undergone root canal therapy and is now infected) could emerge as an active interference field. He identified the two most serious life stressors as pregnancy and influenza.[2(p375)] He also noted that "overloads" to the body could increase the seriousness of infection and of interference fields. Overloads include malnutrition, use of drugs and alcohol, and many forms of toxicity—the same types of things than can block successful treatment.

There are primary, secondary, and even tertiary interference fields. Secondary and tertiary fields are caused when an initial (primary) interference field is left untreated for a significant period. The original interference field causes disturbances elsewhere in the body, to the point that the distant areas also become interference fields. Williams explained the situation in the following way:

> *An appendix scar interference field often causes a disturbed field in the lumbar vertebrae and sacroiliac joints, with resulting low back pain. Over time, these chronically irritated joints can begin to act like a self-perpetuating focus themselves, and . . . may then begin to affect the brain. This newly created brain disturbed field can cause symptoms of fatigue, depression, memory loss or headaches.*[1(p421)]

The existence of primary, secondary, and tertiary interference fields explains the disappearance of many symptoms when one interference field is treated. According to Alfred Pischinger, author of *The Extracellular Matrix and Ground Regulation*, an illness with a single interference field is rare.[3(p102)] The existence of secondary and tertiary interference fields also explains why treating secondary and tertiary interference fields has little effect on symptoms. Treating these types of interference fields leaves many links in the chain intact, and the disturbance is perpetually re-created. Conversely, treating the primary interference field has a domino effect, whereby all secondary and tertiary interference fields are simultaneously treated, along with their distant manifestations of pain and discomfort. Notably, many subclinical interference fields are treated when a primary interference field is addressed. This is the reason for performing Manual Energy Evaluation at a distance of 18 to 24 inches: to identify *only* the primary interference field.

Deciding which interference field to treat

With practice, any of the methods discussed in Chapter 7 will identify interference fields. It is then up to the practitioner to decide which interference field is the primary one in need of treatment. For the practitioner who is new to energy medicine, finding the primary interference field may seem a challenging task. However, most new practitioners discover—to their surprise—that it is not as difficult as they imagined.

For those who are trained in Autonomic Response Testing (ART) or another form of kinesiology, specific methods can be used to determine which field is the primary interference field. Kinesiology connects the practitioner (through the autonomic response system) to the patient's body, allowing the body to "tell" the practitioner which interference field to treat first. Likewise, as the practitioner becomes proficient at Manual Energy Evaluation (MEE), he or she will recognize that this method can be even easier and much faster than ART. Typically, the largest vector or indentation or the hottest or coldest spot is the primary source of disturbance. If you are using the CAMS device in nonflashing mode to identify interference fields, the fields that produce the strongest buzz or vibration in your hand will be the primary fields to treat. Intuition is one of the best guides there is.

For the practitioner who is not yet comfortable with any of the methods mentioned in the previous chapter, there are two good places to start.

1. **Treating the point of pain**

   In some cases, treating the point of pain will resolve an interference field. This is especially true if the injury is recent. Reestablishing normal cellular membrane electrical potential can resolve inflammation and its accompanying pain. This is why

CAMS devices are an ideal adjunct to any sports medicine program. However, as already discussed, an interference field often causes symptoms in a distant area of the body. In these cases, treating the area where the symptoms appear will not solve the problem, although it will often reduce inflammation and pain. The offending interference field must still be found and treated for symptoms to subside completely.

2. **Treating a spot for unresolved emotional conflict**
   As noted in Chapter 6, many, if not most, physical symptoms have an emotional component. In such cases, regardless of the circumstances that caused development of the emotional component, there is one interference field that will consistently be present. This interference field is located on the top of the head, above and behind the right ear (see diagram in Chapter 6). This interference field is often the primary interference field and the link to secondary and tertiary fields elsewhere in the body. Treating this interference field *first* can thus resolve many seemingly unrelated difficulties in distant areas of the body. In my own clinical experience, treatment of this interference field can resolve up to 80% of *all* presenting problems. In other words, if a patient has a list of symptoms, treatment of this one interference field can often resolve the majority of them. Remember, however, that an interference field may need to be treated many times (daily or several times a week) to retrain the body's bioenergy field and to bring lasting coherence.

Being specific

Practitioners who are new to energy medicine often ask about treating the entire bioenergy field. Although this may seem like a good idea, experience indicates that a focused approach is the best way to achieve reproducible results. This is demonstrated by the fact that treating a whole scar is often not specific enough. Sometimes, only specific areas of a scar (evidenced by redness, tenderness, and/or unusual scar tissue) are causing the interference field; the whole scar is not involved. In such cases, treatment should be focused on the specific area. The more specific the treatment, the grea cess. The same is true for other kinds of interference fie ample, differentiating between an interference field in a tooth and an interference field in the tonsils is much more effective than treating the entire mouth. Likewise, identifying one vertebra is usually much more effective than treating the whole lower back. For example, the fingertip placed on a specific spot can help to localize the primary problem; the medical term is not required (Figure 8A).

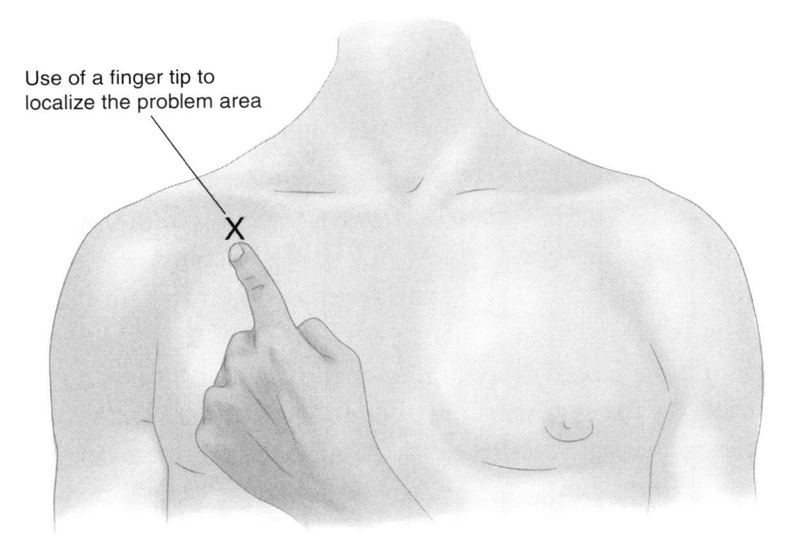

Figure 8A *Use of the finger to localize the primary problem*

Using the CAMS devices

Once the practitioner has determined which interference field to treat, it is a simple matter of turning on the CAMS device and following the directions provided. Each unit is slightly different, they all operate according to the same general protocol:

1. Activate the flash mode.
2. Hold the unit 18 to 24 inches from the selected interference field.

3. Think of the problem area with healing intention.
4. Take a breath and exhale sharply through your nose (the equivalent of a short snort). Do not exhale through the mouth.
5. Hold the unit in place for the appropriate treatment time, which varies with the unit. Most units have a built-in timer.

Intention is a powerful part of the process to maximize the potential of the CAMS devices (see Chapter 5). Marcel Vogel's appointed successor, Rumi Da, explained it this way: "The breath coheres the energy and gives it direction, projecting it to whatever location is expressed in the intent" (personal communication, January 2010). For some, exhaling (snorting) may feel a bit uncomfortable at first, but the exhalation need not be obvious or loud—most patients do not notice a quick nasal exhalation.

Selecting duration and frequency of treatment

With the CAMS technology, more is not necessarily better. The idea is to bring coherence to the interference field and to re-establish normal membrane electrical potential at the cellular level, providing a jump start. Treatment helps the body to "remember" and to readjust to its former, "normal" level of function. The body must then be allowed to hold the new level on its own for as long as possible. It is a matter of retraining the tissues.

In my own clinical experience, overtreatment can actually nullify the effects of treatment—like holding a baby's hand too long when she is ready to walk. Holding the baby's hand for too long creates the impression that the baby cannot walk alone. Similarly, real-time ultrasound observations of CAMS treatments have shown that the beneficial effects of a treatment usually reach a peak at 4 minutes (author's personal observations). Longer treatment may actually reverse the healing effects. Most of the CAMS devices have a timer that is set to treat for 2 minutes or 10 seconds, depending on the device model. This is typically all that is required. Following the initial treatment, a

quick retest or rescan will indicate whether an addition
or 10-second treatment is necessary. More than this would usually constitute overtreatment and is not typically necessary during a single session. However, there are exceptions to this general rule.

If the interference field has existed for many years, and if the diseased condition has progressed to severe levels, longer treatment times may be indicated. The severity of the condition is also an indicator of how many treatments may be required. Regular treatment (daily or several times a week) may be necessary to resolve long-term conditions. Conversely, if development of the interference field is recent, one or two treatments many be all that are required. When the patient feels the results, he or she should determine when to return for further treatment, on the basis of recurrence of symptoms.

Generally speaking, several interference fields can be treated during a single session. One advantage of the CAMS technology is the complete absence of contraindications—there is no way to cause harm. However, CAMS devices should not be used during pregnancy.

Treating pets and plants

Harold Burr's work with living fields—what he called L-fields—indicated that all life forms have a surrounding energy field and that this field holds the patterns for healthy growth and development.[4] In our modern environment, the toxins and pollutants that affect humans also have effects on pets and plants. These organisms can develop anticoherent areas in their bioenergy fields, just as humans do, and these areas of anticoherence can be effectively treated with the CAMS technology.

Pets

MEE is helpful in determining an animal's interference fields, as it is for identifying human interference fields. The CAMS device can be used in nonflashing mode for the same purpose. Treatment follows the same protocol as for humans and can be performed while the animal is either moving or resting. Several veterinarians have discovered the CAMS technology and are using it with great success. L. E. Beltran, BVetMed, PhD, MRCVS, of Ottawa, Ontario, Canada, comments:

> *I have been using the PulseCam unit for about 18 months. I have been a veterinarian for 38 years and have often been thrilled by new approaches and new treatment protocols. The most recent use of the PulseCam has been revolutionary in adding a new and deeper form of pain management but more importantly in tissue/organ healing. I have had spectacular response to the treatment of very painful gallbladder distention pain, acute hydronephrosis pain, and multiple patients presenting with deep myofascial trigger point pain. I have not seen such a novel approach work so effectively, particularly in patients that are not influenced by "placebo effect." I can now not think of working without having the PulseCam available for those patients with deep myofascial pain/injuries. The PulseCam is replacing my deep block injections and IMS [intramuscular stimulation] techniques. This is amazing technology.*

Plants

Over the past several years, the citrus trees in Florida have suffered severe and increasing damage from a disease called citrus greening disease (also known as Huanglongbing or HLB). This bacterial disease, which is spread by an insect, results in small, bitter-tasting fruit, which is unmarketable, and early death of the tree. In 2009, the CAMS technology was used in a preliminary study of a citrus grove infected with the HLB-causing organism. Each week during the growing season, the trees in one row were treated for 2 minutes each. The entire energy field of each tree, including the foliage and the root ball, was treated by means of a sweeping motion with the CAMS device. At the end of the season, the treated row exhibited exceptional, unblemished fruit of a significantly larger size than elsewhere in the grove. Besides being larger and unblemished, these fruits had better flavor, as reported by many of the workers who participated in the study (author's unpublished research).

This preliminary work with plants opens a broad vista of possibility. Further studies are now under way in Florida. The use of scalar energy coupled with intention, as combined in the CAMS technology, has limitless, untapped potential in the agricultural industry.

References

1. Williams L. *Radical medicine.* International Medical Arts Publishing; 2007.
2. Williams L. *Radical medicine.* International Medical Arts Publishing; 2007. Citing Price W. *Dental infections: oral and systemic.* Vol. 1. Penton Press Company; 1923. p. 268–274.
3. Pischinger A. *The extracellular matrix and ground regulation: basis for a holistic biological medicine.* Eibl I, translator. North Atlantic Books; 2007.
4. Burr H. *The fields of life.* Ballantine Books; 1972.

Part II

Conditions That Respond to CAMS Treatment

According to Peter Dosch, author of the most widely read German reference manual on neural therapy:

> *Interference fields upset the economy of all the vital processes and send out their pathogenic impulses via the neurovegetative system [connective tissue network], which is present everywhere in the body. These impulses can manifest themselves at any point of the body as chronic disorder, with symptoms that may be those of painful rheumatism or neuralgia [stabbing or throbbing pain], of circulatory or metabolic disturbances, or of some functional disorder that can ultimately provoke organic changes.*[1(p114)]

Thus, the conditions that may result from an interference field are almost limitless, and any chronic ailment may be due to an interference field. Dosch specifically mentioned circulatory dysfunctions (such as heart disease, blood pressure problems, and arterial disease), metabolic disorders (such as diabetes mellitus and diseases of the liver and kidney), and functional disorders in which the symptoms result from psychological factors, emotional conflicts, or stress.

At first, it may seem presumptuous to suggest that so many different ailments can be relieved by the treatment of interference fields. Yet ultimately, all disorders have one thing in common: anticoherence in the bioenergy field. Given our current understanding of the connection between the bioenergy field, the connective tissue matrix (the communication system within the body), and the cell membrane, it becomes easier to comprehend the influence of such disturbances.

All disorders have one thing in common: anticoherence in the bioenergy field.

Dietrich Klinghardt, MD, PhD, who at the time of writing was president of the American Academy of Neural Therapy, has specified a broad spectrum of conditions that respond to the treatment of interference fields[2]:

- allergies and asthma
- arthritis
- back pain and whiplash
- bladder dysfunctions
- chronic pain
- colitis and ulcers
- depression
- dizziness
- ear problems
- emphysema
- glaucoma and inflammatory eye disease
- headaches, including migraines
- heart disease and circulatory disorders
- hormone imbalance
- kidney and gallbladder disease

- liver disease
- menstrual cramps
- muscle and sports injuries
- prostate disorders
- sinusitis
- skin diseases
- thyroid dysfunction

Each of these conditions is a prime candidate for CAMS treatment. Part II of this volume discusses in more detail the use of CAMS therapy in the treatment of several of these conditions.

References

1. Dosch JP, Dosch M. *Manual of neural therapy according to Huneke.* 2nd English ed. (translation of 14th German ed.). Gutberlet R, translator. Thieme Medical Publishers; 2007.
2. Trivieri L, Anderson J, editors. *Alternative medicine: the definitive guide.* 2nd ed. Celestial Arts Publishing Co; 2002. Quoting Klinghardt D.

Chapter 9

Chronic Pain, Inflammation, and the CAMS Blue Light

Pain is classified in two major categories: acute and chronic. Acute pain refers to any pain occurring immediately after injury; it functions as a warning that something is wrong. According to Western medicine, chronic pain is different; it serves no apparent purpose, but neither does it respond to the medical model of care.[1] Chronic pain is usually defined by the medical profession as any unexplained pain that continues for longer than would be expected for normal healing to occur.

By 2002, chronic pain had become the single most common health disorder in the United States, affecting 86 million Americans and costing nearly $90 billion a year in medical bills and lost wages. The problem has not gone away, and chronic pain is currently the most frequently cited reason for disability.[2]

With no apparent cause and no apparent cure for chronic pain, the medical profession is left with the all-too-familiar refrain, "It's just something you'll have to learn to live with." This phrase is often accompanied by another familiar phrase: "Try this pain medication, and if it doesn't work we'll try something else."

Chronic pain often affects areas of the body that are distant from one another, and it can be difficult to trace them without an understanding of the ways in which connective tissue and acupuncture meridians unite the organs and systems of the body. Those who understand the nature and development of interference fields realize that chronic pain *does* have a cause: anticoherence. More importantly, chronic pain is often easily resolved by treating that cause, by restoring order and coherence to the body's energetic field. This, in turn, restores order to the internal connective tissue matrix and ultimately to each cell.

It is interesting that some of the most successful alternative modalities for the treatment of chronic pain (transcutaneous electrical nerve stimulation [TENS], acupuncture, and bodywork)[2(p663–664)] all address the health and condition of the connective tissue matrix. These methods can effect changes in the cell membrane's electrical potential, with positive effects on at least some kinds of chronic pain. The CAMS technology goes one step further, addressing chronic pain at the level of its underlying cause within the bioenergy field. Witnessing, by means of ultrasound technology, the return of normal cell membrane electrical potential and the restoration of connective tissue in real time has astounded many practitioners. Speaking of this phenomenon, Charles Schwengel, DO, MD(H), director of the Rhythm of Life clinic in Mesa, Arizona, had this to say:

> *With ultrasound images of injured or inflamed areas on a monitor, we can watch in real time the reduction and resorption of extracellular fluids and shrinkage of [scar] tissue. The first time we observed this, the very experienced and competent technician's jaw dropped and he said, "That just doesn't happen!"*

Treatment of interference fields has become one of the most popular ways to deal with chronic pain in Germany.[3] In a study conducted in the 1970s, the pain of trigeminal neuralgia (severe facial pain) was either completely eliminated or substantially improved in more than 70% of the 639 cases treated.[4] Compared with Western medicine's poor track record with this type of chronic pain (which includes the use of muscle relaxants, anticonvulsant medications, pain medication, and surgery),[5] the potential benefits are obvious.

The inflammatory component of chronic pain

One of the most significant components of chronic pain is inflammation. Many people recognize inflammation for its short-lived symptoms: swelling, redness, heat, pain, and decreased range of motion. However, when the inflammatory response does not shut down properly (which is typical when an interference field is involved), inflammation continues at a low level. This low-level inflammation can play a pivotal role in the persistence of chronic pain and in the gradual deterioration of many areas of the body.[6(p101–102)]

The medical community has become increasingly aware that inflammation underlies some of the most devastating afflictions of our time: heart disease, cancer, diabetes, arthritis, chronic fatigue, osteoporosis, multiple sclerosis, fibromyalgia, thyroid dysfunction, Alzheimer disease, Crohn disease, and many others. The evidence is so strong that in 2003 the American Heart Association and the Centers for Disease Control and Prevention recommended that doctors include a test for inflammation in their medical check-ups to determine patients' risk for heart disease.[7]

The reason that the connection between inflammation and disease has gone unnoticed for so long is that this type of inflammation is just below the usual threshold of detection. It can be present for years without any symptoms, all the while producing stress and dysregulation in the connective tissue matrix.[6(p101–102)] This is one reason for the term "silent inflammation," which is used to describe the low-level, symptomless inflammation that underlies so many conditions. According to Alfred Pischinger, author of *The Extracellular Matrix and Ground Regulation,* silent, chronic inflammations cause energy-consuming limitations and tissue acidosis.[6(p180)] Pischinger further states, "If this now unstable system receives a minor stimulus or gets a secondary noxious exposure, there will be an inadequate and excessive response to this 'second blow' which will set off a distant disturbance."[6(p102)]

Chronic pain, silent inflammation, and abnormal cellular electrical potential all have a common root: interference fields. According to Dosch, in his classic text on neural therapy, there is a direct connection between inflammation and cell membrane electrical potential. He says, "A precondition for any inflammation and for many other pathological processes is a change in the electrical potential of the cell membrane."[8(p78)] James Oschman noted, in his foreword to Pischinger's book, how easy it can often be to restore normalcy: "Subtle stimulation at the appropriate points can elicit the release of long-held toxins and resolution of so-called silent inflammation."[9(p xiv)] CAMS treatments represent a form of subtle stimulation. Directed at appropriate points, such treatment can elicit the resolution of silent inflammation and the subsequent relief of pain.

Types of chronic pain

It is not within the scope of this book to provide details on the various types of chronic pain. That information is available from many other sources, including the Internet. However, an overview of the main categories of chronic pain is included here to give a sense of the potential of CAMS treatment as it relates to pain.

Visceral pain: Visceral pain occurs within the main body cavity and generally corresponds to an internal organ. The pain itself often feels like a deep ache, sometimes with cramping. Visceral pain may radiate to other locations, as with heart disease, which may involve pain radiating to the shoulders and neck. Other examples of visceral pain include irritable bowel pain, bladder pain and cystitis, prostate pain, and the pain of endometriosis.

Somatic pain: Somatic pain pertains to the framework and musculature of the body, as distinct from the viscera. It is often characterized as a sharp pain localized in a specific area. Somatic pain occurs in the skin and muscles, as well as the joints, bones, and ligaments. Examples include headaches, arthritis, and fibromyalgia.

Neuropathic pain: Neuropathic pain is caused by damage to the nerves or the spinal region. It includes symptoms such as numbness, burning, and tingling (as with diabetic neuropathy), but also sharp stabbing pains. Examples include sciatica, trigeminal neuralgia, and carpal tunnel syndrome.

Psychogenic pain: Psychogenic pain, also called psychalgia, is physical pain that is caused or prolonged by emotional or behavioral factors. Examples include headaches, back pain, and stomach aches. These types of pain are candidates for treatment of the emotional interference field (see Chapter 6).

A 54-year-old female school teacher described her experience with the CAMS technology for treatment of pain as follows:

> *Three months ago, I developed an oral herpes outbreak and Bell's palsy [paralysis of one side of the face]. I never suffered so much pain in my life. The sores in and out of my mouth were very painful, and the temporary paralysis was very uncomfortable. I was given medication to help control the symptoms and the pain. Most of this medicine did not work. Using TensCam a few minutes once a day, I felt a great deal of relief from the nagging pain on the side of my face and in my ear. It also helped to alleviate my inability to close my left eye and straighten my mouth so I could smile normally. I used the TensCam for about three weeks. I believe it was responsible for much of my progress.*

Whether the pain is acute (with corresponding acute inflammation) or chronic (often with accompanying silent inflammation), the CAMS technology can often have an almost immediate effect. Even acute pain is accompanied by an interference field that typically resolves on its own. Gentle encouragement, by re-establishing coherence in the interference field, can greatly speed recovery. In the case of acute injuries, CAMS treatment directed *at the point of pain* rapidly normalizes cell membrane electrical potential with an almost resolution of inflammation and pain. The overall healing be greatly accelerated. This is illustrated by the following two case histories shared by Mark Orbay of the Family Naturopathic Clinic in Ottawa, Ontario, Canada:

> *A 46-year-old male sustained a knee injury while ice skating. One week later when I saw him, the knee was still somewhat swollen and painful when he put any kind of pressure on it. One treatment with the TensCam device brought 70% improvement within 3 days. From that point on, he continued to improve until he regained full use of his knee.*
>
> *An 88-year-old male dropped a heavy flower pot on his foot. He went to the hospital and the X-rays revealed no fracture. A couple of days later he hobbled into my office with a cane. When I examined him, his foot was severely bruised, swollen, and very painful to the touch. During treatment with the TensCam device he felt a vibratory sensation—a good sign of stimulation in the inflamed tissues. I also prescribed natural medicines for pain relief. At his one week follow-up appointment, the bruise had completely disappeared, the inflammation had receded, and he was walking normally without a cane.*

The resolution of chronic pain usually requires the location and treatment of a primary interference field. Although chronic pain may require more treatments for full resolution, the same process takes place: normalization of cell membrane electrical potential, reduction of silent inflammation, and resolution of pain. Simon Trueblood, MD, an anesthesiologist and pain management specialist in Merrillville, Indiana, had this to say of using the CAMS technology for treatment of pain:

> *The TensCam unit is the most impressive device I have used in pain management. It has obviated the need for invasive interventional techniques in many instances.*

Inflammation and microbial infection

Inflammation often begins with microbial infection. Although the symptoms of infection may eventually subside (indicating that the infection has resolved), the microbes may persist at low levels—just enough to disrupt normal cell membrane electrical potential. Under conditions of stress and trauma, infections of this nature often migrate to other areas of the body, where they can cause noticeable symptoms and further infection. Sometimes, the secondary infection also remains symptomless for many years, accompanied by silent inflammation. Ill health, toxic exposure, or any form of additional stress can cause full-blown infection, inflammation, and the development of chronic pain.

Microbes (bacteria, viruses, algae, fungi, and yeasts) may also hide in what are known as biofilms, slimy, gel-like substances produced as a protective shield by organized colonies of microbes. Microbial biofilms develop into complex structures that resemble miniature cities, with channels for the flow of nutrients and the excretion of wastes. Biofilms are virtually impenetrable by antibiotics and other chemical methods intended to eradicate them. They escape traditional methods of treatment and become resistant forms of infection. However, biofilms are more vulnerable to energetic methods of eradication like the application of certain electromagnetic wavelengths.

Recent research has demonstrated that blue light at a wavelength of 470 nanometers (nm) effectively controls many types of bacteria, including periodontal bacteria, the bacteria that cause acne,[10] and drug-resistant *Staphylococcus aureus* (also known as MRSA).[11] The 470-nm wavelength of visible light is effective in controlling both gram-positive and gram-negative bacterial strains.[11]

The newest innovation in CAMS technology includes blue lights that resonate at the 470-nm wavelength. When the CAMS device is held 1/2 to 3/4 inch from the skin, various types of bacteria and many other microbes can be eliminated. This is a valuable adjunct to treatment of interference fields and helps to relieve the body of a microbial burden that is otherwise very difficult to eliminate. The area should be treated for 2 minutes every 4 to 6 hours for 3 or 4 days. In many instances, such therapy can replace oral antibiotics, thereby avoiding their adverse effects and their cost.

References

1. Singh M, Patel J, Gallagher RM. Chronic pain syndrome. In: eMedicine. Medscape; 2009 Jun. Available from: *http://emedicine.medscape.com/article/310834-overview*
2. Trivieri L, Anderson J, editors. *Alternative medicine: the definitive guide.* 2nd ed. Celestial Arts Publishing Co; 2002.
3. Klinghardt D, Wolfe B. Presentation. Advanced Neural Therapy Workshop; 1992 Dec 5–6; Santa Fe (NM).
4. Trivieri L, Anderson J, editors. *Alternative medicine: the definitive guide.* 2nd ed. Celestial Arts Publishing Co; 2002. p. 667. Citing Dosch P. *Facts about neural therapy according to Huneke: regulation therapy—brief summary for patients.* Medicina Biologica; 1985.
5. Trigeminal neuralgia. Mayo Clinic. Available from: *http://www.mayoclinic.org/trigeminal-neuralgia/*
6. Pischinger A. *The extracellular matrix and ground regulation: basis for a holistic biological medicine.* Eibl I, translator. North Atlantic Books; 2007.
7. Pearson TA, Mensah GA, Hong Y, Smith SC Jr. CDC/AHA Workshop on Markers of Inflammation and Cardiovascular Disease. CDC/AHA workshop on markers of inflammation and cardiovascular disease: application to clinical and public health practice: overview. *Circulation* 2004;110:e543-e44. Available from: *http://circ.ahajournals.org/cgi/content/full/110/25/e543*
8. Dosch JP, Dosch M. *Manual of neural therapy according to Huneke.* 2nd English ed. (translation of 14th German ed.). Gutberlet R, translator. Thieme Medical Publishers; 2007.
9. Oschman J. Foreword. In: *The extracellular matrix and ground regulation: basis for a holistic biological medicine.* Eibl I, translator. North Atlantic Books; 2007.

10. Morton CA, Scholefield RD, Whitehurst C, Birch J. An open study to determine the efficacy of blue light in the treatment of mild to moderate acne. *J Dermatolog Treat.* 2005;16(4):219–223.

11. Enwemeka CS, Williams D, Hollosi S, Yens D. Blue light photo-destroys methicillin resistant *Staphylococcus aureus* (MRSA) in-vitro. In: Waynant R, Tata DB, editors. *Proceedings of Light-Activated Tissue Regeneration and Therapy Conference.* Springer US; 2008. p. 33–37. Available from: *http://www.springerlink.com/content/g531t6r776l86462*

Chapter 10

Musculoskeletal Disorders

The musculoskeletal system, which consists of the bones, joints, muscles, and connective tissue, has a cause-and-effect relationship with the entire body. When the skeletal components are misaligned, muscles are stretched or pulled to compensate, and the blood supply and the signals conducted by the nervous system are impeded. A good example of this situation is apparent with imbalances that originate as discrepancies in leg length. Misalignments of this nature can throw off the hips, forcing the spine to compensate. This in turn causes the shoulder muscles at the base of the skull to shorten, the tissues at the back of the neck to become stretched, and the jaw to become contracted—all of which place increased pressure on the nerves along the base of the skull and down the neck and shoulders. Besides the stress and headaches that can develop, other functions of the body may be affected, including the digestive, immune, hormonal, and even brain functions. In particular, misalignments often reduce the supply of blood to the brain. All of these outcomes may result from an interference field caused by a childhood fall, tension due to emotional stress, or any number of other factors. Treatment of the appropriate interference field can sometimes correct misalignments without physical force. John E. Haugland, DC, of Winter Park,

Florida, had the following to say of the CAMS te culoskeletal problems:

> *I have found the TensCam to be fast, easy, an[illegible] than other equipment for my musculoskeletal patient practice.*

The increasing prevalence of musculoskeletal disorders has been described as an epidemic. As with any condition, the presence of toxins, malnutrition, and stress add to the burden and exacerbate both the onset and the severity of symptoms. Among the most common musculoskeletal problems are back pain, many forms of arthritis and other inflammatory conditions, and fibromyalgia or chronic fatigue syndrome. Injuries and muscle cramps also fall into this category. Each of these categories is addressed briefly below.

Back pain

Back pain is the second leading cause for visits to a physician in the United States, and low-back pain is the third most common reason for surgery.[1] According to Doug Lewis, ND, past chair of the Physical Medicine Department of Bastyr University in Kenmore, Washington, "The site of [back] pain is rarely the site of the dysfunction. You may make the pain subside, but you're not correcting the dysfunction that caused the pain in the first place. If you leave the pain alone and treat the cause, then you have the pain as a monitor for whether or not your therapy is working.[2]

Many forms of alternative therapy for back pain follow Lewis's advice. Some of the most effective alternative therapies aim to normalize cell membrane electrical potential. Two of these are neural therapy (discussed in Chapter 6) and its cousin, prolotherapy (the injection of a proliferant such as dextrose, along with a local anesthetic, into the area of pain), which was successfully used by the US Surgeon General C. Everett Koop, MD, for over 20 years for ᴇ treatment of back pain. Like neural therapy, prolotherapy re-establishes normal membrane electrical potential in the area of pain while stimulating the body's natural healing process.

Another alternative therapy that supports cell membrane electrical potential is treatment with a TENS device (transcutaneous electrical nerve stimulator). TENS devices supply a small electrical current to the affected area. This modality is believed to stimulate endorphins, which are natural pain blockers. However, because the devices use electrical current, they also have the capacity to stabilize the electrical potential in the surrounding cells, and this may be a secondary explanation for the relief provided.

Acupuncture (especially electro-acupuncture) is another treatment method that affects cell membrane electrical potential. The main objective of acupuncture is to unblock the meridian pathway to the area of discomfort. With the normalization of cell membrane electrical potential, cellular communication to and within the area of discomfort is re-established, while inflammation and pain are allowed to subside.

Many forms of bodywork, including massage, osteopathy, and chiropractic, relax muscles, restore structural orientation, and release toxins. These methods also help to re-establish normal cellular electrical potential. Even exercise goes a long way toward the restoration of normal membrane potential, since movement opens many channels in the body.

Each of these forms of therapy for back pain support normalization of the tissues in the area of pain and/or misalignment. When the area of pain is itself the primary interference field, these methods are successful. But when the interference field is elsewhere in the body, long-term relief is seldom established. The fact that TENS devices and other therapies typically provide only temporary relief of pain indicates that the offending interference field is not being treated.

The missing piece in understanding many forms of back pain (and other musculoskeletal conditions) is the presence of a distant interference field. The displacement of a bone or joint never occurs in isolation. The soft tissues connected to the bones determine where the bones "sit." When conditions in these tissues are out of balance (i.e., when there is an interference field), the corresponding skeletal components often become misaligned. Treatment of the area surrounding the dysfunctional bone or joint can correct the position and the movement of a joint. Misalignments often come into alignment during CAMS treatment—without physical manipulation.

A 75-year-old retired US Navy captain commented on relief of his back pain:

> *I suffered from chronic back pain for many years and often sought relief from chiropractic treatments. Although I was a skeptic, I applied the TensCam to my chronic back pain. To my surprise, after a few applications, the pain subsided and I've not needed outside help for about two years. When I sense the pain returning, two or three TensCam treatments keep the problem under control. I've found similar amazing results with an arthritic elbow and an old football injury to my knee. All things considered, I'm a skeptic-turned-believer.*

Arthritis

The letters "itis" at the end of a medical term refer to inflammation of a particular part of the body. The word "arthritis" means joint inflammation. Although joint inflammation describes a symptom rather than a specific condition, the term "arthritis" is often used to refer to any disorder affecting the joints, including osteoarthritis, gout, rheumatoid arthritis, infectious arthritis, and ankylosing spondylitis.

Arthritis has many "accepted causes," including toxins, microbes, hormonal factors, genetic predisposition, age, allergens, and dental infection. The fact that arthritis can be initiated by so many different factors is a prime indicator of its deeper cause—an interference field. The presence of an interference field weakens the body and impedes its communication network to the area. When a person is in a weakened state, any number of different agents may lead to the ɔpment of arthritis. Louisa Williams has pointed out that an ...erference field in the tonsils (either from chronic infection or from tonsillectomy) can cause arthritis or heart disease later in life.[3(p379)] Weston A. Price and others demonstrated that interference fields in the teeth could cause arthritis and other degenerative diseases.[4] Even stress can cause hormonal imbalance resulting in arthritis.[5]

Arthritic disorders fall within the broader category of rheumatic diseases, which are characterized by inflammation and subsequent loss of function. Because one of the underlying issues with arthritis is inflammation, and because inflammation is so rapidly addressed by treatment with the CAMS technology, arthritis is often relieved by treating the painful area or areas. In my own practice, I have frequently observed the reduction of swelling and the return of mobility with treatments of 2–4 minutes. However, longer-term relief can be gained when the offending interference field is found and treated. Often, treatment of the primary interference field will resolve arthritic pain in many different areas of the body. The following case history, reported by Norman Smith, DC, and Paul Campbell, ND, of Pineville, Missouri, illustrates how effective CAMS treatment can be for arthritic conditions:

> *The 54-year-old male patient presented with bone-on-bone osteoarthritis in his left knee and was anticipating knee replacement. He also had degenerative joint disease in the lower cervical and lower lumbar spine, antalgic posture and a painful limp. A four-minute CAMS treatment to the knee caused all pain to disappear. Further two minutes on neck and lower back with same results. Two weeks later, still no pain and evidence that regeneration is taking place.*

Another illustration of relief of arthritis comes from patient Pete Reilly:

> *For quite a few years I've experienced considerable arthritic pain in both hands and feet, especially my feet. My family doctor*

prescribes pain medication. The medication does little good. A couple of minutes with the TensCam unit really is a life saver. It is not a permanent one-time fix, unfortunately, but it sure is great.

Tendonitis

Tendonitis is inflammation of a tendon, the fibrous tissue that connects muscles to bones. The body has hundreds of tendons, but tendonitis tends to occur in only a few of them. The most common places for tendonitis are the shoulder, knee, and wrist. Tendons in these areas have a propensity toward reduced blood supply, especially with increased use. Irritation in an area where the supply of blood is already reduced leads to disturbances in cell membrane electrical potential, which in turn leads to inflammation and tissue damage.

Most treatments for tendonitis include anti-inflammatory medication and other measures to reduce the inflammation. However, as long as the inflammation persists, the membrane electrical potential cannot return to normal, and the cycle of tissue disturbance will continue. Because CAMS treatments address disturbances in membrane electrical potential, they are very effective in bringing the inflammatory cycle to a close. Many practitioners treating tendonitis have noted the obvious signs of reduced inflammation (changes in skin temperature and color and reduction of swelling) during CAMS treatment.

Bursitis

A bursa is a fluid-filled sac around a joint that normally promotes smooth, pain-free motion. Bursitis is inflammation of the bursae. It results from repetitive motion or pressure on particular bursae, for example, kneeling or continuously leaning on the elbow. [illegible]
an inflammation, it can often be resolved quickly with C[illegible]
ment directed at the point of pain.

Fibromyalgia and chronic fatigue syndrome

Fibromyalgia and chronic fatigue are syndromes rather than diseases. This means that each of these conditions is a collection of signs, symptoms, and medical problems that tend to occur together but that are not related to a specific, identifiable cause. Fibromyalgia and chronic fatigue syndrome share many of the same signs and symptoms: muscle and joint pain, sleep disorders, digestive problems, and severe fatigue. Some specialists actually consider these conditions together as "severe chronic fatigue states."

According to the website of the National Institute of Arthritis and Musculoskeletal and Skin Diseases, many people associate the development of fibromyalgia with a physically or emotionally stressful or traumatic event.[6] Leon Chaitow, ND, DO, author of more than 65 books including *Fibromyalgia Syndrome: A Practitioner's Guide to Treatment*, has also noted that both fibromyalgia and chronic fatigue syndrome often begin after an infection or a severe trauma.[7(p15)]

In the case of emotional trauma, the obvious spot for treatment with CAMS technology is the interference field corresponding to unresolved emotional conflict: the place at the top of the head, above and behind the right ear (see Chapter 6). Treatment of trigger points can often ease localized pain in minutes.

Injuries

The musculoskeletal system is affected by a number of other conditions, including injuries (strains, sprains, bruises, breaks, etc.). The CAMS technology is a must for the treatment of sports injuries. Early treatment re-establishes the integrity of the bioenergy field, resolves cellular membrane electrical potential, and reduces inflammation so that normal healing can occur. Mark Orbay, a naturopathic physician in Ottawa, Ontario, Canada, shares this case history, which illustrates how treatment soon after an injury often results in an immediate response:

> *A 60-year-old female tripped at the entrance of her apartment. To break the fall, she fell on her out-stretched hands. When I*

saw her four hours after the accident, both wrists were stiff and swollen. She was in excruciating pain. Minutes after a treatment with the TensCam, her pain disappeared. By the next morning she reported the swelling dissipated and she regained full range of motion and complete use of her wrists.

Muscle cramps

The skeletal muscles in the calf, thigh, and arch of the foot are the most notorious sites of muscle cramps. Such cramps occur when the muscles involuntarily contract and then do not relax. They are more common in older people and in athletes, who put greater-than-normal strain on certain muscles. Although a medical explanation for cramps has been difficult to pinpoint, electrolyte imbalance is often considered a major factor. Exercise and age both affect electrolytes. Given what is known about electrolytes and membrane electrical potential, it is safe to assume that normalizing the electrical potential at the cellular level will help to alleviate cramps. This is often the case, as illustrated by the experience of a scuba diver from South Africa:

In September 2009, my daughter and I had the privilege to go on a week-long scuba diving excursion to the Great Barrier Reef. Our arrangement allowed us to dive four or five times a day. Unfortunately during one of the early dives, my daughter and I each started having leg cramps. Not only are leg cramps painful, they also place a diver at risk for a serious diving accident. We started drinking tonic water for its quinine conte[illegible] a desperate attempt to end the leg cramps. W[illegible] benefit from the quinine water but w[illegible] about the safety factor. We fear[illegible] was about to end.

Then a fellow diver became aw[illegible] to let us use his CAMS Personal Tun[illegible] lem. We accepted his offer and allow[illegible] run for two minutes, directed at ea[illegible] minutes to see the results of the treatment. To my delight, the cramps were gone and have not returned. My daughter, 25 years my junior, had the same problem and was helped in the same way. We went diving for several more days and had no more problems with leg cramps. Subsequently, I returned to South Africa and have had no further difficulty. I related this experience to my family physician who became very interested and is planning to purchase a TensCam for her medical practice.

Menstrual cramps

During menstrual periods, a woman's uterus contracts to help expel the uterine lining. Prostaglandins, hormone-like substances involved in pain and inflammation, trigger these contractions. Many experts believe that severe contractions constrict the blood vessels feeding the uterus. The resulting pain is comparable to angina, which occurs when blocked coronary arteries deprive portions of the heart of nutrients and oxygen.[8] When cramps result from inflammation and/or from restricted blood flow, the CAMS technology can be extremely helpful. Simply directing the device at the lower abdomen can relieve symptoms for hours (see also Chapter 13).

References

1. Trivieri L, Anderson J, editors. *Alternative medicine: the definitive guide.* 2nd ed. Celestial Arts Publishing Co; 2002.
2. Trivieri L, Anderson J, editors. *Alternative medicine: the definitive guide.* 2nd ed. Celestial Arts Publishing Co; 2002. Quoting Lewis D.
3. Williams L. *Radical medicine.* International Medical Arts Publishing; 2007. p. 379.

Williams L. *Radical medicine.* International Medical Arts Publishing; 2007. p. 388. Citing Price W. *Dental infections: oral and systemic.* Vol. 1. nton Press Company; 1923.

ri L, Anderson J, editors. *Alternative medicine: the definitive* nd ed. Celestial Arts Publishing Co; 2002. p. 535. Citing Peat

R. Hormone balancing: natural treatment. *J Rheum Dis Med Assoc.* 1986;1(1).

6. Fibromyalgia. National Institute of Arthritis and Musculoskeletal and Skin Diseases; 2009. Available from: *http://www.niams.nih.gov/Health_Info/Fibromyalgia/default.asp*
7. Chaitow L. *Fibromyalgia syndrome: a practitioner's guide to treatment.* 3rd ed. Elsevier; 2009.
8. Menstrual cramps: causes. Mayo Clinic; 2009. Available from: *http://www.mayoclinic.com/health/menstrual-cramps/DS00506/DSECTION=causes*

Chapter 11

Circulatory Disorders

Almost everything finds its way into the circulatory system of the body—not only nutrients and other beneficial substances, but also infectious microbes, toxins, and other harmful substances, which are carried through the system for dispersal or elimination. Weakness in any area along the way is an invitation for microbes to take up residence, for toxins to build up, and/or for silent inflammation to progress. In time, these secondary or tertiary sites of infection or toxic buildup can develop into interference fields.

Many, if not all, forms of heart disease are accompanied by inflammation, as are arterial diseases. Even hypertension (high blood pressure) has an inflammatory component, which is due to the buildup of plaque in constricted arteries. This is why inflammation is considered a risk factor for heart disease. The microenvironment of the heart favors microbial growth.[1(p369)] Beyond initiating many forms of heart disease, infection and inflammation can result in valvular damage and may manifest as heart murmurs.

According to Peter Dosch, interference fields cause about 30% of all cardiac disorders.[2(p179)] This represents a huge number of patients, considering that heart disease is the number one cause of death in the United States. Yet it is possible that Dosch has underestimated the percentage. The heart is often not the primary interference field, but typically becomes a secondary or tertiary interference field as a result of a disturbance elsewhere in the body. For example, it is well known that rheumatic fever, an immune response to the presence of streptococcal bacteria, can precipitate valvular heart disease.

Not as well understood is the acupuncture meridian connection between the wisdom teeth and the heart. The wisdom teeth (the third molars) have a direct association with both the heart and the small intestines. Many holistic dentists and physicians are aware of the serious impact that wisdom teeth, and the infections harbored in their cavitation sites, can have on cardiovascular function.[1(p518)] Christopher Hussar, an osteopathic doctor, dentist, and leading authority on cavitation surgery, believes that incorrect extraction of wisdom teeth and the resulting low-level infection represents "the worst global disease in the world."[3]

Microbes have what is called a pleomorphic quality, being able to change their shape, size, and pathogenicity according to their environment.[1(p369)] The microbes that commonly reside in tonsil and dental interference fields are often of low virulence. However, under the right conditions (e.g., stress, toxicity, trauma), they migrate and manifest in another form, with greater pathogenicity, in other areas of the body. This explains how a symptomless interference field caused by subclinical infection in one area of the body can cause a symptomatic secondary interference field in another area of the body. Thus, when searching for the primary interference field related to heart disease, the tonsils and the wisdom teeth are good places to start. Additionally, when the tonsils or any of the teeth are implicated, it is helpful to use the CAMS blue light to treat the infection itself (see Chapter 9).

The following case history reported by Norman Smith, DC, and Paul Campbell, ND, of Pineville, Missouri, illustrates the resolution of congestive heart failure and other issues related to the same interference field.

The 49-year-old female patient was diagnosed with congestive heart failure and presented with a hard, painful, walnut-sized lump on the left breast. After two, two-minute CAMS treatments, the breast lump stopped hurting and was reduced to pea-size. Four days later, it was completely gone. All signs of congestive heart failure are now gone and confirmed by cardiologist after three CAMS treatments in three weeks.

Other circulatory disorders that may be correlated with interference fields include arrhythmias (disorders of the heart's rhythm), diseases of the arteries (including atherosclerosis), and valvular diseases, each of which is discussed briefly below.

Arrhythmia

Arrhythmia, also referred to as dysrhythmia or heart palpations, refers to abnormal electrical activity in the heart. The heartbeat may be too fast or too slow, or it may be irregular. Some forms of arrhythmia are life-threatening and may result in cardiac arrest and death. Others cause less severe symptoms, such as infrequent palpitations. Still others may not be associated with any symptoms at all, but they may predispose the patient to potentially life-threatening stroke.

The most recent research, published in 2009, has revealed the psychological and emotional component of many arrhythmias, particularly unresolved anger.[4] Anger tests are now being considered as predictors of future arrhythmia.[4] Because the heart is recognized in Traditional Chinese Medicine as the seat of emotion, one place to look for an interference field associated with arrhythmia is the emotional spot on the top of the head, as described in Chapter 6.

As described in Chapter 1, Valerie Hunt, PhD, of UCLA witnessed the resolution of dysrhythmia early in her work with treatments of the energy field. She recorded the frequency pattern of a dysrhythmic heart while simultaneously recording the disturbed pattern in the bioenergetic field. When the coherency of the energy field was improved, the anticoherent pattern in the heart dissipated. At the physical level, breathing was normalized and the heart itself came under automatic control.[5(p247)]

Atherosclerosis

Atherosclerosis occurs when the inner walls of the arteries are narrowed by a buildup of plaque, which consists of cholesterol, cellular waste products, calcium, and other substances. Plaque can grow to the point that it significantly reduces blood flow, but most of the damage occurs when the plaque ruptures, causing blood clots.

For patients with atherosclerosis, the critical factor is not the presence of cholesterol, but the presence of chronic inflammation. Long-term inflammation destabilizes cholesterol deposits in the coronary arteries, which increases the risk that a portion of the plaque will dislodge. The resolution of chronic inflammation by means of the CAMS technology can have a significant effect on atherosclerosis and many other arterial diseases. As noted earlier, direct treatment of the heart is not usually necessary; the offending interference field is often found elsewhere.

A Note of Caution: CAMS treatment is not recommended for patients with cardiac pacemakers of the "demand" or "sensing" type. The CAMS device may inhibit or otherwise interfere with the actions of such pacemakers. As a precaution, patients with known heart disease who are considering treatment with a CAMS device should undergo evaluation by a physician to determine if any risk factors are present.

Valvular diseases

In a normally functioning heart, the four valves (flaps) keep blood flowing in the appropriate direction at the right time. They act as gates, swinging open to allow blood to flow through, then shutting tightly until the next cycle begins. Valvular heart disease refers to any dysfunction of one or more of the heart's four valves: the mitral and aortic valves on the left side and the tricuspid and pulmonic valves on the right side.

According to the American Heart Association's heart and stroke statistics update for 2010, valvular heart disease is responsible for more than 21,000 deaths each year in the United States and is a contributing factor in about 44,000 deaths.[6] The majority of these cases involve disorders of the aortic and mitral valves.

Health professionals from many different disciplines acknowledge the innate ability of the body to heal once it is unencumbered by toxins and other limiting factors. This book and the CAMS technology are about the greatest of these limiting factors—the condition of the surrounding energy field. When distorted energetic patterns are released from the bioenergy field, the body can accomplish even seemingly impossible feats. This capacity is illustrated by two case histories in which valvular disorders were resolved as the energy field was treated. Richard Armond, an osteopathic physician in Suwanee, Georgia, reports:

> *In 2002, my mother had a heart attack—one of several she had experienced over the years. On this occasion, doctors told her that the murmur (a leaky mitral valve), was much worse. There was nothing the conventional medical profession could offer. I had just been introduced to the CAMS technology and I drove eight hours to Florida to treat her in the hospital. I made several weekend trips and treated her both osteopathically as well as with the TensCam. Surprisingly, she regained energy and returned home. Subsequent tests by her cardiologist revealed that the leaky valve was improving. Ultimately, the murmur disappeared and my mother lived another 6 years—till just before her 92nd birthday.*
>
> *Not too long after my mother's recovery I had another opportunity to treat a valvular heart disorder. This time, the patient was a newborn who had been diagnosed with pulmonic stenosis. Stenosis is a tightening of the valve where blood flow is severely impeded. Doctors informed the parents that the boy's condition would likely worsen and require a procedure to break open the valve before he was a year old. The parents brought him to my office for regular osteopathic treatments. With each*

treatment, I also used the TensCam. The pulmonic function gradually improved and the pediatric cardiologists suggested rechecking at yearly intervals.

References

1. Williams L. *Radical medicine*. International Medical Arts Publishing; 2007.
2. Dosch JP, Dosch M. *Manual of neural therapy according to Huneke*. 2nd English ed. (translation of 14th German ed.). Gutberlet R, translator. Thieme Medical Publishers; 2007.
3. Williams L. *Radical medicine*. International Medical Arts Publishing; 2007. p. 519. Citing Shane R, Williams L. *A manual of dominant focus therapuetics*. Self-published; 1993. p. 10.
4. Lampert R, Shusterman V, Burg M, et al. Anger-induced T-wave alternans predicts future ventricular arrhythmias in patients with implantable cardioverter-defibrillators. *J Am Coll Cardiol*. 2009;53(9):774–778.
5. Hunt V. *Infinite mind: science of the human vibrations of consciousness*. Malibu Publishing Co.; 1996.
6. Lloyd-Jones D, Adams RJ, Brown TM, et al. Heart disease and stroke statistics 2010 update: a report from the American Heart Association. *Circulation*. 2010;121(7):e46–e215.

Chapter 12

Gastrointestinal Disorders

The liver, gallbladder, stomach, pancreas, and intestines form a single functional unit for the digestion and assimilation of nutrients. The health of the gastrointestinal system is paramount to the health of the whole organism. If one of these organs becomes compromised, the whole body suffers from malnutrition to some degree. This can, in turn, be the stress that triggers the development or activation of an interference field elsewhere in the body.

Conversely, a disorder of the gastrointestinal system may be triggered by an interference field in any part of the body. In such cases, the organs themselves can develop into secondary or tertiary interference fields. When the gastrointestinal system is involved, treating all of the interference fields in the chain is an especially good idea. Treatment of the primary field stops the outgoing message of disharmony, while treatment of the affected organs reduces inflammation and resets the membrane electrical potential in each area to allow proper healing to occur.

The organs of the gastrointestinal system are often closely associated with emotional issues. For example, Traditional Chinese Medicine connects the stomach with worry; the liver with anger, frustration, and depression; the gallbladder with resentment and bitterness; the intestines with lonesomeness; and the pancreas with issues of self-worth. The emotional connection should never be overlooked. A discussion of some common gastrointestinal problems that can be supported and/or resolved with CAMS treatment follows.

Reflux

Gastroesophageal reflux disease, commonly referred to as GERD or acid reflux, is a condition in which the contents of the stomach back up or reflux into the esophagus. GERD is considered by the medical profession to be a chronic condition—once it begins, it is supposedly incurable. Even after apparently "successful" medical treatment, the problem usually returns within a few months. Individuals with GERD often end up taking antacids and other medications for the rest of their lives. This ultimately leads to other problems.

According to Robert Kidd, reflux is almost always accompanied by an interference field at the gastroesophageal junction, where the esophagus connects to the stomach.[1(p60)] Treatment of this area may reduce or eliminate symptoms entirely. In fact, treatment at this point is effective for the relief of many gastrointestinal symptoms, including ulcers and gastritis.[2(p114)] It is often successfully combined with treatment of other interference fields associated with organs of the gastrointestinal tract.

Gastritis

Gastritis is a common problem defined as inflammation of the lining of the stomach. It has a variety of symptoms, including abdominal pain, nausea, indigestion, and loss of appetite. Up to 10% of those who visit hospital emergency departments reporting abdominal pain have gastritis. Given that gastritis often results from the regular use of nonsteroidal anti-inflammatory drugs (also known as NSAIDS, a drug class that includes aspirin, ibuprofen, and naproxen) the high frequency of gastritis may be related to the fact that NSAIDS are taken daily by a growing number of individuals.

Treatment of the gastroesophageal junction often resolves the primary interference field, and treatment of the stomach itself reduces inflammation so that the healing process can proceed.

Ulcers

As with most gastrointestinal conditions, duodenal and gastric ulcers are inflammatory disorders. Regardless of the causative agent, such disorders generally will not heal until the inflammation is brought under control. CAMS treatment is extremely helpful for this purpose because it is able to reduce inflammation and allow healing to begin. The subsequent location and treatment of the primary interference field stops the chain of events that is sending disharmony to the area.

Gallstones

About 40% of all women over 40 years of age have gallstones, and this condition is particularly common among those who have borne children.[3(p130)] In many patients, the symptoms of gallstones manifest after the birth of a child. This suggests the presence of an interference field, for which surgery is not the answer.[3(p131)] Perhaps this is why more than 30% of gallbladder surgeries result in no improvement of symptoms. For gallbladder problems experienced by women who have had children, Peter Dosch, author of *Manual of Neural Therapy According to Huneke*, suggests treatment of an interference field in the pelvic region.[3(p131)]

One of the interesting things about gallstones is that 90% of patients who have them never experience symptoms. This is another indication that gallbladder symptoms result from interference fields. Without a trigger (such as stress or trauma), most people with gallstones never have problems. Sometimes, the gallbladder itself is where treatment should be focused. This was the case for a woman whose treatment was monitored by ultrasonography, which showed the complete resolution of inflammation during 1 minute of treatment with the CAMS device. Her story and frames from the monitoring video are presented here.

> *When I found out about CAMS treatments, I had already spent $9,000 during a weekend hospital stay. All their tests revealed no apparent cause for the severe symptoms I was experiencing. Dr. Crosby immediately identified my gallbladder as the problem and he hooked up an ultrasound scanner to verify. Then we watched the scanner during treatment. Even I could recognize the changes on the ultrasound viewer during the one minute treatment. My symptoms vanished and I have not had any trouble for over ten years.*

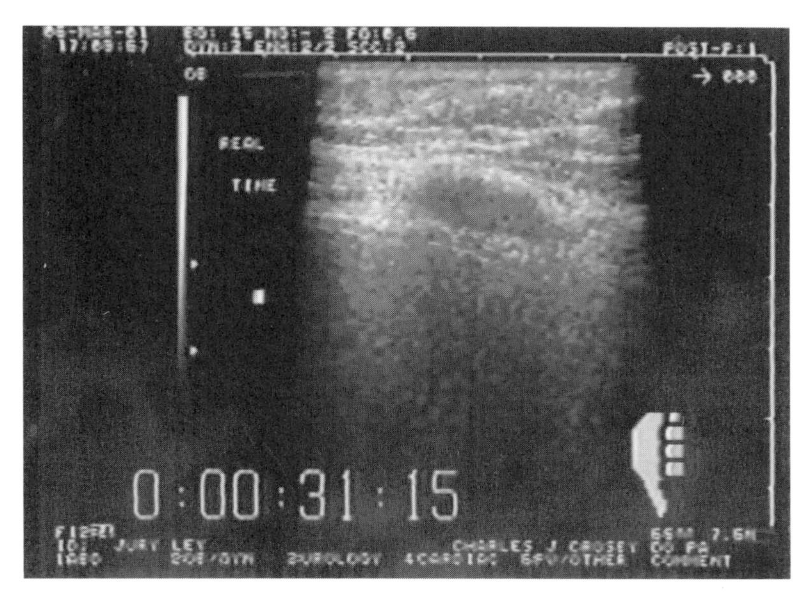

1. Before PulseCam treatment: *The dark area indicated by the arrow shows the edema of the patient's inflamed gallbladder.*

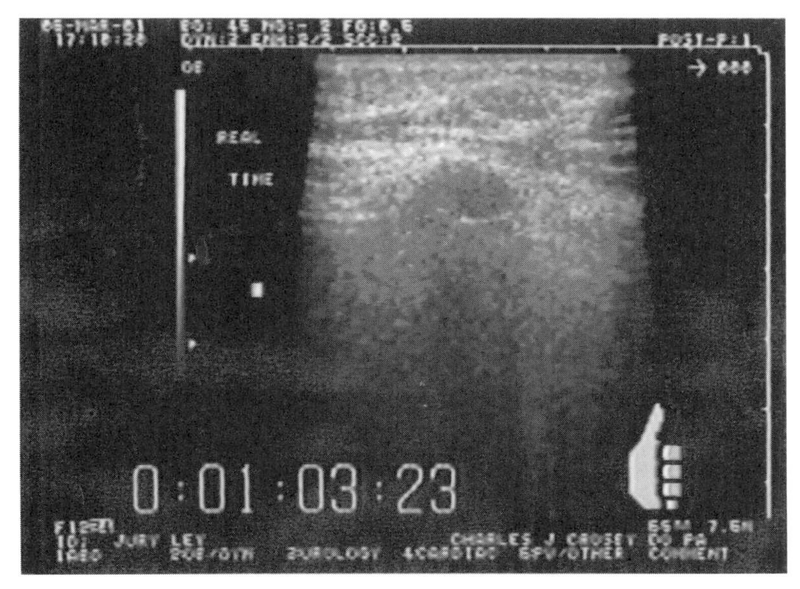

2. After 32 seconds of PulseCam treatment: *The dark area is beginning to get smaller.*

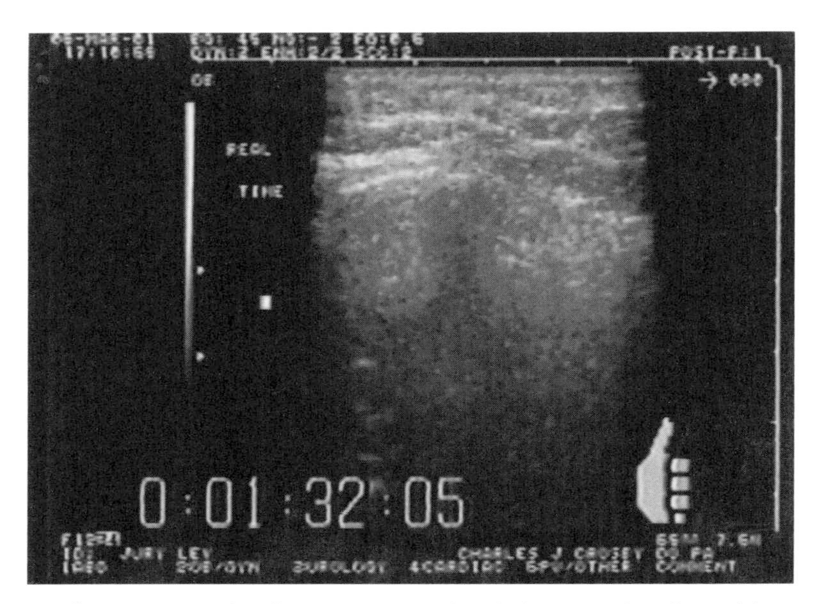

3. After 61 seconds of treatment: *The dark area is barely visible.*

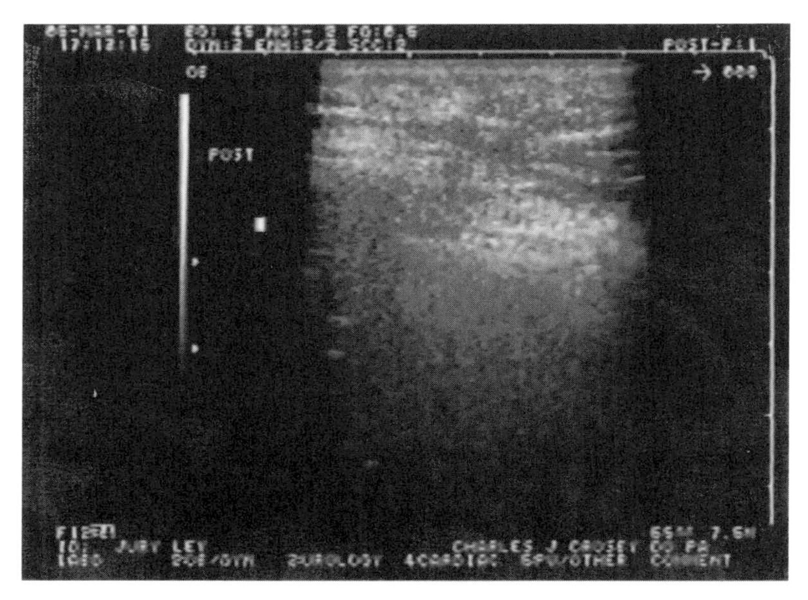

4. The final picture after 1 minute: *The edema has disappeared. The patient simultaneously reported that her pain was gone. Ten years later, the patient remained asymptomatic.*

Liver interference field

The liver is quite possibly the most overworked organ in the human body. It bears the burden of removing the huge number of toxins to which we are exposed in our modern environment. The liver is responsible for detoxifying the blood and for removing all the pollutants we encounter, be they from air, water, or food. The liver must also break down drugs (both over-the-counter and prescription) before their elimination from the body. When a person has been taking medication for an extended period, the liver often becomes an interference field. In such cases, the body may be overwhelmed with toxins, a circumstance that can nullify CAMS treatment. Because the liver is also connected with anger, one of the most toxic of emotions, the burden on this organ is increasingly heavy if the patient also has unresolved anger.

In Traditional Chinese Medicine, the liver is also associated with depression. As such, toxins in the environment may be linked, through the liver, with depression. Mild toxic exposures do not usually cause problems for healthy people. However, those who have experienced serious liver stresses, such as long-term use of drugs (over-the-counter or prescription), hepatitis, or mononucleosis, are more vulnerable. Past stresses of this nature are likely to have left an inactive interference field. Further stress can activate the field and cause the sudden onset of symptoms.

For a variety of reasons, an overworked liver often becomes a primary interference field. Not only can the presence of a liver interference field influence the digestive process but it can also result in depression, apathy, and irritability. It may also be responsible for referred pain, particularly to the right shoulder. Louisa Williams, in her book *Radical Medicine*, pointed out the viscerosomatic connection between the liver, the spinal nerves of the upper and mid back, and pain in the right shoulder and arm.[4(p410)] This type of pain, which does not respond to structural therapy, may be due to a primary interference field in the liver.

Treatment of a primary interference field in the liver can, in some cases, provide instant improvement in pain, mood, and energy level. In such cases, the "boost" provided by treatment is enough to re-establish normal function. More often, repeat treatments are required. In the case of a liver interference field, any program of detoxification will support the healing process.

Inflammatory bowel diseases

The term "inflammatory bowel disease," often referred to as IBD, covers a group of disorders in which the intestines become inflamed. The three most common types of inflammatory bowel disease are ulcerative colitis, Crohn disease, and irritable bowel syndrome. There is little doubt that stress affects the digestive system. All of the inflammatory bowel diseases are associated with and aggravated by stress. Additionally, many drugs, including the NSAIDS, are associated with inflammation of the gastrointestinal organs.

Treatment at the point of pain is often the best place to start.[3(p154)] This will help to resolve the inflammation and will go a long way toward supporting the healing process. Location of the primary interference field, which may be at the esophageal junction (as mentioned above) or which may be associated with some unresolved emotional conflict, will complement the treatment of the point of pain. Resolution of irritable bowel syndrome by this means often resolves chronic low-back pain at the same time.[4(p846)]

Kidd has noted that chronic intestinal infections often initiate an interference field in the celiac plexus, the nerve ganglion located between the sternum and the belly button. This is the nerve control center for the entire gastrointestinal tract. For any patient who has had an acute gastrointestinal illness (parasitic, yeast, or bacterial), especially if there has not been a full recovery, the possibility of an interference field in this area should be considered.[1(p29,32)]

Heartburn

Heartburn, a burning sensation in the upper chest that may come in waves about an hour after eating, is a daily occurrence for about 10% of Americans.[5] It is considered a precursor to the more serious condition of GERD or reflux, described above. Heartburn occurs in people with a weakened esophageal valve. Pressure from the stomach causes the weakened valve to allow the acidic contents of the stomach to enter the esophagus. Many people are able to control heartburn by eating less, especially in the evening, which reduces the pressure on the esophageal valve.

As with GERD, CAMS treatment of the junction between the esophagus and the stomach often helps to relieve heartburn. If the heartburn is infrequent, one or two CAMS treatments may be sufficient to resolve the problem. For those with serious heartburn (which may, in fact, be undiagnosed GERD), several treatments are typically required. Evidence indicates that CAMS treatment supports the strengthening of the esophageal valve for the more permanent resolution of heartburn and GERD.

Hernia

A hernia is an opening or a weakness in the muscular wall of some cavity of the body. The easiest way to visualize a hernia is to imagine an enclosure with a hole or weakness and a balloon that is blown up inside the enclosure. Once the balloon is inflated, it will bulge out through the hole or weakness. In this illustration, the hole represents the hernia, which allows tissues to bulge out from their normal position within the body. There are many kinds of hernias, but some of the most common occur within the abdomen. Activities that can worsen a hernia are lifting, coughing, or even straining to have a bowel movement.

The main concern with hernias is the possibility that tissues may become trapped, a process called incarceration. Such trapped tissues may become cut off from the blood supply, and surgery may be required.

As noted in the discussion of heartburn, CAMS treatment has been used to strengthen weakened tissues in the body, including the esophageal valve and the valves of the heart (see discussion of valvular diseases in Chapter 11). Similarly, treatment of herniated tissues appears to strengthen the tissue and resolve the hernia in many cases. Hernias have been reported to close without surgery, as in this example, reported by a fellow doctor:

> *I had a hernia the size of a hen's egg and was not interested in having it surgically repaired. After four treatments with the TensCam (once a week for four weeks), the size of the hernia was reduced to the size of a pencil eraser.*

Hemorrhoids

Hemorrhoids are another example of tissues that become weakened and swollen because of pressure. In this case, the tissues are the veins in the anal and rectal region. Hemorrhoids result from pressure caused by a variety of digestive and lifestyle factors: diarrhea, constipation, fiber-deficient diet, dehydration, excess weight, hypertension, inappropriate posture, heavy lifting, and, for women, pregnancy. In each of these circumstances, the veins in the rectal and anal areas experience extra pressure, which results in swelling accompanied by itching and burning sensations.

The treatment of hemorrhoids with neural therapy involves injections of procaine around the anus.[6] CAMS treatment, which is noninvasive and therefore much more respective of the patient's privacy, is as simple as directing the CAMS device at the hemorrhoidal tissue from a distance of 18 to 24 inches. As with the treatment of hernias, strengthening of the tissue can sometimes resolve the problem. Several treatments may be necessary.

References

1. Kidd R. *Neural therapy: applied neurophysiology and other topics.* Custom Printers; 2005.
2. Dosch M. *Atlas of neural therapy with local anesthetics.* 2nd English ed. (translation of 5th German ed.). Lindsay A, Grossman J, translators. Thieme Medical Publishers; 2003.
3. Dosch JP, Dosch M. *Manual of neural therapy according to Huneke.* 2nd English ed. (translation of 14th German ed.). Gutberlet R, translator. Thieme Medical Publishers; 2007.
4. Williams L. *Radical medicine.* International Medical Arts Publishing; 2007.
5. Heartburn/GERD guide. Web MD, LLC. Available from: *http://www.webmd.com/heartburn-gerd/guide/understanding-heartburn-basics*
6. Kidd R. The pelvic floor. *Neural Ther Pract* [electronic newsletter]. 2006 Aug;1(5). Available from: *http://www.neuraltherapybook.com/newsletters/1-5.php*

Chapter 13

Genitourinary Disorders

The genitourinary system consists of the organs and structures of the urinary and reproductive systems in both males and females. It is responsible for a variety of functions, including the production and regulation of hormones, reproduction, and the filtering and excretion of wastes. The genitourinary system interacts and works with every other system in the body. It is subject to a variety of disorders that may disrupt the patient's everyday life, including bladder, kidney, prostate, and menstrual disorders.

Bladder incontinence

Incontinence affects about one-third of the female population.[1] The term "incontinence" is often used to describe a group of symptoms, including irritable bladder, overactive bladder, painful bladder syndrome, interstitial cystitis, and urge incontinence. Any of these related conditions may be due to an interference field in the bladder itself or to an interference field in the kidneys.[2] Treatment over the bladder or over the kidneys is often all that is necessary to relieve the symptoms.

The osteopathic profession has long recognized bladder irritability as being related to somatic dysfunction, whereby the misalignment of the pubic bone puts pressure on the urethra.[2] Osteopathic manipulation to straighten the pubic bone often releases the pressure and, consequently, the symptoms. CAMS treatment over the pubic bone may allow the body to naturally correct this misalignment on its own, solving the problem without manipulative intervention. CAMS treatment may also resolve inflammation, for even quicker resolution of symptoms.

Electrostimulation by means of transcutaneous electrical nerve stimulation (TENS) has been effective in relieving the symptoms of interstitial cystitis.[3] Robert Kidd has noted the similarity between TENS and the CAMS technology for the treatment of this condition:

> *It is interesting that electrostimulation using a variety of techniques and frequencies, (from 5 to 50 Hz) has been shown to be effective in a large number of [cystitis] trials. I cannot help but notice the resemblance to neural therapy, using the Tenscam device. The Tenscam delivers an 8 Hz "energy" (neither electrical nor magnetic) and seems to have similar effects to that of procaine. So it would seem that the electrostimulation is working in a similar way to neural therapy i.e. modulating and regulating the autonomic nervous system.*[2]

Traditional Chinese Medicine links bladder and kidney problems with the emotions of fear and shock. When the patient's history indicates the presence of an emotional connection of this type, treatment of the location on the top of the head, above and behind the right ear (as discussed in Chapter 6), should be considered, in addition to treatment of the bladder and urethra. As with other conditions, an interference field located anywhere may cause problems of the genitourinary system.

A Note of Caution: Bladder distress may also involve infection or even kidney stones (discussed in the next section), two possibilities that should not be overlooked. If painful urination and/or blood in the urine persists longer than 24 hours after CAMS treatment, the patient should seek further medical attention.

Kidney stones

The passing of a kidney stone can be one of the most painful events a person will ever experience. The process may cause intense pain in the back, side, abdomen, groin, or genital area, and it may be accompanied by nausea. The spike-like projections of a kidney stone also cause inflammation and swelling as it passes down the ureter. The medical profession offers little help, other than pain medication during the process, which may take up to 3 weeks. If the kidney stone is unusually large or the patient has multiple stones, a treatment known as shock wave lithotripsy may be used to break up the stones into pieces of a more manageable (and less painful) size for elimination. This procedure usually requires a hospital stay of several days.

As a noninvasive alternative to strong medication and/or admission to hospital, the CAMS technology represents a huge step forward for the treatment of kidney stones. Depending on the size of the kidney stone, the reduction of inflammation and swelling is often enough to allow the stone to pass freely without distress. Some people report freedom from recurrent kidney stones following CAMS treatment. My own personal experience is a good example of both:

> *In 2005, I woke at 4 a.m. with acute, lower abdominal pain which I recognized as classic kidney stone symptoms. I treated the area with a TensCam for several minutes until the pain subsided. At 9 a.m. I visited a general practitioner who confirmed the kidney stone diagnosis (pain pattern and blood in the urine). I had a busy schedule planned for the day and I decided to proceed with my day's activities. Throughout the day, each time the pain returned, I treated it with the TensCam. I was able to*

control the pain all day. At the end of the day, another urine sample confirmed no blood—indicating the passage of the stone. I never experienced any more pain and never had any further symptoms.

Prostatitis

Like irritable bladder, prostatitis is a group of disorders with related symptoms that can cause pain in the groin, painful urination, difficulty urinating, and more frequent urination, especially at night. A multitude of factors may be involved, including disorders of the immune or nervous system, infection, stress, pressure on the prostate from other diseased tissue, and traumatic injury, as might be caused by riding a horse or a bicycle.

Regardless of the apparent cause, prostatitis is a manifestation of an autonomic regulatory disturbance and is the result of an interference field.[4(p226)] Treatment of the disturbed field improves the blood supply to the area, stimulates tissue nutrition, and mobilizes the body's healing power. In a German study, treatment of interference fields located at the prostate in 400 patients with prostatitis resulted in relief of symptoms in 92% of cases. Hypertrophy (enlargement of the prostate) was resolved in 80% of cases.[4(p226)]

Menstrual pain

As many as 60% of all women experience menstrual cramps.[5(p121)] This type of pain occurs as the uterine muscles contract to expel tissue and fluids during menstruation. The congestion of blood and other fluids causes inflammation, with resulting cramping and pain. The most common treatment for menstrual pain is anti-inflammatory drugs, which work well for short-term relief. However, CAMS treatments work faster and the effect can last longer, without any adverse effects. Treatments directed to the lower abdomen and pubic area reduce inflammation, which improves blood flow and relaxes the muscles. Other symptoms accompanying the menstrual cycle, such as bloating, headache, diarrhea, and water retention, also often subside. Sometimes, CAMS treatments even resolve the structural issues that underlie menstrual pain.

Menstrual problems are also recognized as having many psychological connections. Both stress and repressed pain from the past can affect the menstrual cycle, resulting in painful periods. In such cases, the practitioner might also consider treating the interference field for unresolved emotional conflict, on the top of the head, above and behind the right ear (as discussed in Chapter 6).

Ovarian cysts

Most ovarian cysts are harmless and in fact are quite normal. However, some cysts cause problems if they fail to regress on their own, within 2–3 months. These persistent cysts may eventually cause pain, bleeding, and other problems, and some may become cancerous. Many ovarian cysts have been treated effectively with CAMS devices directed at the ovaries or at the point of discomfort. Correcting disturbances in the energy field can hasten the normal resolution of many cysts and often circumvents more severe consequences down the road.

Pregnancy

As a precaution, CAMS devices should not be used over the abdominal area, if at all, during pregnancy.

References

1. Female incontinence: what you should know and why you should tell your doctor. *Ebony* 2005 Aug. Available from: *http://findarticles.com/p/articles/mi_m1077/is_10_60/ai_n14858990/*
2. Kidd R. Bladder interference fields. *Neural Ther Pract* [electronic newsletter]. 2009 May;4(5). Available from: *http://www.neuraltherapybook.com/newsletters/4-5.php*

3. Interstitial cystitis/painful bladder syndrome. NIH Publ. No. 10–3220. National Kidney and Urologic Diseases Information Clearing House; 2009. Available from: *http://kidney.niddk.nih.gov/kudiseases/pubs/interstitialcystitis/*
4. Dosch JP, Dosch M. *Manual of neural therapy according to Huneke*. 2nd English ed. (translation of 14th German ed.). Gutberlet R, translator. Thieme Medical Publishers; 2007.
5. Northrup C. *Women's bodies, women's wisdom: creating physical and emotional health and healing*. Revised and updated. Bantam Books; 1994.

Chapter 14

Neurological Disorders

Neurological disorders affect the nerves of the brain and spinal cord, as well as the peripheral nerves. Symptoms may be as simple as muscle weakness and tingling or as serious as facial drooping, numbness, slurred speech, and paralysis. Neurological conditions are often extremely painful, because of nerve damage resulting from infection, inflammation, injury, or disease. For patients with neurological conditions, the medical profession often has little to offer. In contrast, CAMS treatments represent an open doorway with many new possibilities—everything from the treatment of anger and depression to the treatment of Reflex Sympathetic Dystrophy, carpal tunnel syndrome, and post-polio syndrome. These and other neurological disorders are discussed in this chapter.

Anger, anxiety, depression, and other mood disorders

Although anger, anxiety, depression, and other mood disorders are often considered emotional or psychological problems, they have a definite link with the nervous system and with the functions of the brain. This connection has been demonstrated in numerous studies, in which electrostimulation of the brain has increased mood-altering neurohormones and neurotransmitters. Levels of various compounds, such as serotonin, gamma-aminobutyric acid (GABA), and dehydroepiandrosterone (DHEA), increase with stimulation of the brain, leading to significant improvement in the symptoms of anxiety, depression, mood disorders, headache, and even insomnia.[1(p88–95),2]

Serotonin is a neurohormone that is known to alter mood. Numerous medications that affect the action of serotonin are used to treat depression. GABA is a neurotransmitter that regulates anxiety and many other cortical functions. DHEA has been referred to as the "youth hormone." Studies have correlated DHEA levels in the blood with a feeling of well-being, as well as with health and vigor. In addition to producing changes in these neurohormones and neurotransmitters, electrostimulation of the brain also decreases the stress hormone cortisol.[1(p88–95)]

CAMS treatment directed to the head appears to have many of the same effects as electrostimulation. A CAMS treatment of 2–4 minutes can significantly reduce anger and lift depression for a majority of patients. Regular treatment has helped numerous individuals to regain normal emotional stability. Other benefits have included improved focus and clarity and enhanced sense of well-being. Treatment should be directed to the entire head.

Neuropathic pain

Neuropathy is a collection of disorders that occur when nerves of the peripheral nervous system (the part of the nervous system outside the brain and spinal cord) are damaged. Neuropathy usually causes pain, weakness, burning, tingling, and numbness. It can eventually result in complete loss of feeling in the extremities. In the United States, about 20 million people suffer from neuropathy, which is initiated by traumatic injuries, infection, metabolic disorders, and, especially, exposure to toxins. Neuropathy can affect the nerves that control muscle movement (motor nerves) and those that detect sensations such as cold or pain (sensory nerves). In some instances, it can affect internal organs, such as the heart, blood vessels, bladder, or intestines, a condition known as autonomic neuropathy.

Two of the most common forms of neuropathy are diabetic neuropathy and chemically induced neuropathy. Diabetic neuropathy, which is experienced by more than half of those with diabetes mellitus, results when higher levels of glucose remain in the blood for long periods of time. Despite the ample supply of glucose in the body, the individual cells cannot receive this metabolic fuel and they become starved for glucose. Long-term blood sugar imbalances result in reduced blood flow to the peripheral nerves, lack of oxygen, free radical damage, and nerve cell death. This damage is usually progressive and irreversible. Other than strict attention to regulation of blood sugar, the medical profession currently has little to offer in terms of halting the progression of neuropathy resulting from diabetes.

Chemically induced neuropathies include those associated with pesticides and other toxic exposures and are usually the result of intermittent exposure over an extended period. Chemotherapy for cancer and other medical conditions is another cause of chemically induced neuropathy.

Neuropathy is a growing problem that can be helped by CAMS treatment, as illustrated by the experience of Don Newburg, a farmer from Iowa who experienced peripheral neuropathy, possibly because of many years of applying pesticides on his farm. Regular CAMS treatment brought his condition under control:

About five years ago, I developed peripheral neuropathy in my feet. I visited the Mayo Clinic and was told there was nothing they could do. I expected to have to take pain medication and to live without feeling in my feet for the rest of my life. On the evening of my first TensCam treatment, I was surprised to be able to feel the cool sensation of the tile as I walked across the floor. I decided to purchase a TensCam so that I could treat myself regularly. Progress was not rapid but within several months (daily treatment), I noticed marked improvement. Today I have no pain, no burning or tingling. I still treat myself almost daily because when I go too long between treatments, I notice that symptoms begin to recur.

An example of relief from neuropathic numbness comes from a 39-year-old woman with multiple sclerosis, as reported by Gregory J. Wastl, DC, of Eagan, Minnesota:

The patient's presenting complaints were bilateral foot numbness with paralysis of the toes, and right hip numbness. Following her first CAMS treatment, right hip numbness was gone and her foot numbness and toe paralysis improved 50%. After her second treatment, the toe paralysis was gone and the foot numbness was barely noticeable.

Nerve entrapment

Nerve entrapment, also known as compression entrapment, is caused by pressure on a nerve, which reduces blood flow and results in inflammation and poor nerve conduction. Nerve entrapment can result in neuropathy, muscle weakness, and complete loss of feeling in the affected part of the body. The condition of the cells of entrapped nerves is similar to the condition of cells with disturbed cell membrane electrical potential. More specifically, inflammation of the entrapped nerves reduces nerve action potential.

The symptoms of nerve entrapment will affect one specific part of the body, depending on which root nerve is affected. For example, symptoms in the arms can result from nerve compression in the neck, shoulder, elbow, or wrist. Symptoms in the legs can result from compression in the back, hip, lower leg, or foot. Treating the *root* nerve with a CAMS device can relieve symptoms rapidly, and regular treatment may resolve the problem entirely.

Carpal tunnel syndrome

Carpal tunnel syndrome is a type of nerve entrapment. The wrist (known in anatomic terms as the "carpus") is surrounded by a band of fibrous tissue that normally functions as a support for the joint. The tight space between this fibrous band and the wrist bones is called the carpal tunnel. It houses the median nerve, which receives sensations from the thumb, index, and middle fingers of the hand. Any condition that causes swelling or a change in position of the tissue within the carpal tunnel can cause inflammation of the median nerve, producing tingling or numbness of fingers or shooting pain in the wrist. This problem can result from repetitive wrist motion (such as typing), weight gain, low thyroid function, vitamin deficiencies, or fluid accumulation and retention.[3] According to Dietrich Klinghardt, MD, PhD, founder of the Klinghardt Academy of Neurobiology, many cases of carpal tunnel syndrome can be traced to interference fields in the arms and shoulders caused by vaccination scars.[4(p1003)] Compression of the nerve root and misalignment of the sixth cervical vertebrae may also be factors.[3]

Carpal tunnel syndrome affects adults of all ages. Women are affected more than men, particularly those who have had a hysterectomy without oophorectomy and those who have gone through menopause between 6 and 12 months previously.[5] These connections speak to the involvement of interference fields elsewhere in the body that should be investigated by the practitioner. However, because carpal tunnel syndrome is an inflammatory condition, treatment of the median nerve in the wrist is a good place to start, and such treatment often brings rapid relief without surgery.

Reflex Sympathetic Dystrophy

Reflex Sympathetic Dystrophy (RSD), otherwise known as complex regional pain syndrome type I, is a chronic nerve disorder that occurs most often in the arms or legs after a minor injury. The damaged nerves are no longer able to control blood flow, sensation, and temperature in the affected area. The symptoms progress from pain and burning with the slightest touch to limitation of muscle movement, contraction of muscles and tendons, severe pain in the entire limb, and muscle wasting. Other typical features of RSD include changes in the color and temperature of the skin, extreme sensitivity of the skin, sweating, and swelling.

The medical profession has little to offer those with RSD, a condition that is considered incurable. Physicians may prescribe topical analgesics, antidepressants, corticosteroids, and opioids to relieve the pain. Other treatments may include sympathetic nerve block, stimulation of the spinal cord, and implantation of drug pumps to deliver pain medication directly to the spinal cord. However, no single drug or combination of drugs has produced consistent, long-lasting improvement.

In my own clinical experience, people with RSD have an energetic leak (i.e., an interference field) in the motor area on the top of the head that corresponds with the affected extremity. This area on the head, referred to as the primary motor cortex or homunculus, represents a motor correlation with individual parts of the body (see Figure 14A).

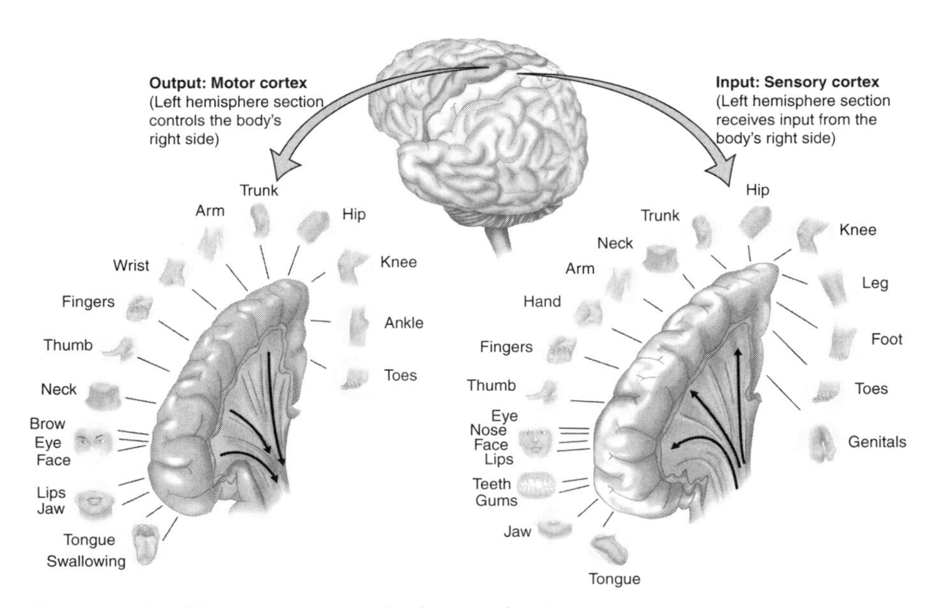

Figure 14A *Homunculus on the human brain*

CAMS treatment should be directed toward the area on the homunculus that corresponds to the affected part of the body (e.g., for pain in the right leg, the CAMS device is directed to the top left portion of the head) and also toward the point of pain, to help rebalance the nervous system and reduce local inflammation.

Canadian physician Ken Boake, shares this case history of using the TensCam to treat RSD:

A 55-year-old woman fell and fractured her left wrist. I saw her three months after the injury at which time the wrist was spastic, numb, and painful when grasping. Obviously, she had Reflex Sympathetic Dystrophy. I offered her a series of treatments with the TensCam. After seven treatments, twice weekly, she said she didn't need any further treatments; everything was back to normal—feeling, color, and movement.

Polio and post-polio syndrome

Polio and post-polio syndrome (PPS) affect the central nervous system, causing muscle weakness and often paralysis. PPS occurs in people who had polio earlier in life and is characterized by new weakening in muscles that were previously affected by the polio infection and in muscles that were seemingly unaffected. In the following example, a woman with PPS recovered her ability to walk after being on crutches for many years. The interference field for this woman was found in the homunculus.

At age five, the presenting woman had contracted the polio virus. It had left one leg severely weakened. In the years that ensued, leg muscle function had deteriorated to the point that crutches were necessary for ambulation. One CAMS treatment directed to the motor area on the top of her head (homunculus), tightened joints and muscles enough so that this woman regained about 80% of leg function and was able to walk without crutches for the first time in many years.

Nerve damage

Patients may experience many kinds of nerve damage. [illegible] nately, the medical profession recognizes few, if any, ways to support or rebuild deteriorated nerve tissue. Experience with the CAMS devices indicates that regeneration can take place when inflammation is controlled and the offending interference field is treated. In the following example, Mark Snyder, a physical therapist, explains how severely damaged nerves were brought to near full recovery with regular CAMS treatment:

> *I was in a serious auto accident in 1989. I had broken ribs and right shoulder injury. The spinal accessory nerve was stretched and almost completely severed. I had no use of my right arm or hand for over a year. I had been a physical therapist for seven years in the 1970s so I understood the importance of keeping my arm and shoulder mobile. I insisted on pool therapy to keep my joints and muscles functional. I was right hand dominant so I enrolled in re-dominance training while I did everything I could to save my right arm and hand. I continued on my own with 12 years of pool therapy. I was able to recover up to 15% of normal grip strength and range of motion. Then, I was in another accident and had ligament damage. I went to an osteopathic doctor who used the TensCam unit directed at the trapezius area. With the first treatment, the trapezius elevated approximately one inch. I was convinced that the TensCam would help. I eventually purchased my own TensCam and I used it faithfully twice a day for two minutes. Within nine months I recovered 80–85% of my arm and shoulder range of motion and grip. My neurologist, MD and PT were all astounded at my recovery. My healing has been permanent and stable for almost 5 years now. I highly recommend the treatment to anyone that has a problem with nerve damage, chronic pain, or similar problems.*

Simon Trueblood, MD, also recognized the ability of the CAMS technology to support nerve health. He used the CAMS devices for many years to help resolve a variety of neurological problems:

> *The CAMS device is the most effective device there is for treating a myriad of problems. Using nothing else, I have successfully treated entrapment neuropathies, complex regional pain syndrome (RSD), headaches, and peripheral neuropathy.*

References

1. Liss S, Liss B. Physiological and therapeutic effects of high frequency electrical pulses. *Integr Psychol Behav Sci.* 1996;31(2):88–95.
2. Shealy N. Depression: a diagnostic, neurochemical profile and therapy with cranial electrical stimulation (CES). *J Neurol Orthopaed Med Surg.* 1989;10(4):319–321.
3. Karpen M. Treating carpal tunnel syndrome. *Altern Complement Ther.* 1995;1(5):284–289.
4. Trivieri L, Anderson J, editors. *Alternative medicine: the definitive guide.* 2nd ed. Celestial Arts Publishing Co; 2002.
5. *The Merck manual of diagnosis and therapy.* 16th ed. Merck, Sharp & Dohme Research Laboratories, Division of Merck & Co., Inc.; 1992.

Chapter 15

Endocrine Disorders

Endocrine disorders encompass a variety of conditions that involve the overproduction or underproduction of hormones. Some of the most common such problems are diabetes mellitus, thyroid disorders, and adrenal disorders. Each of these can be helped by means of CAMS treatments.

Diabetes mellitus

As the seventh leading cause of death in the United States,[1(p391)] diabetes affects a growing proportion of the population. It is considered both a metabolic disease, because it inhibits the absorption (metabolism) of sugar, and an endocrine disorder, because it involves the production and utilization of the hormone insulin.

The pancreas produces insulin, which is essential for the proper metabolism of sugar. Without insulin in the cells, glucose remains in the blood and cannot serve its nutritive purpose. Diabetic individuals are unable to produce *enough* insulin or their cells have become *resistant* to insulin and are thus unable to bring the insulin inside. Either way, excess sugar accumulates in the blood, which reduces the oxygen-carrying capacity of the blood and deprives cells of their source of energy. Lack of insulin or insulin function ultimately causes cell death. What most people do not realize is that insulin is also the body's primary fat-storage hormone, and too much insulin triggers weight gain.

For cases in which insufficient insulin is produced, regular CAMS treatment of the pancreas appears to stimulate the balanced production of insulin. Many of those who receive CAMS treatments have been able to lower their insulin therapy by as much as half. Treatment should be focused over the pancreas (about 4 inches below the left nipple, at the edge of the rib cage).

In cases of insulin resistance, the pancreas produces more than enough insulin; it is accessing the insulin that is the problem. Peter Dosch, author of the *Manual of Neural Therapy According to Huneke,* has found that insulin resistance can be overcome by *repolarizing* the cells.[2(p158)] This observation implies that insulin resistance is related more to membrane electrical potential than to faulty insulin receptors. If the diabetes has not progressed too far, treatment of the offending interference fields can sometimes re-establish normal cellular electrical potential, which in turn restores insulin sensitivity through normal functioning of the insulin receptors.

Kidney failure and chronic kidney disease

Diabetes often leads to kidney failure and the need for dialysis. In fact, diabetes is a contributing factor in about 35% of all cases of chronic kidney disease.[3] High blood pressure causes another 30% of cases of kidney disease. Other causes of kidney disease include infections, narrowing of the renal artery, heavy metal poisoning, and long-term use of over-the-counter pain medications such as acetaminophen and ibuprofen. According to the medical profession, there is no cure for chronic kidney disease. Once it develops, it progresses until dialysis or transplant is required.

When the blood vessels in the kidneys are damaged, the kidneys are no longer able to cleanse the blood. This results in chronic kidney disease, which is typically accompanied by weight gain and ankle swelling because the body is unable to rid itself of wastes. For those with diabetes, high blood sugar levels gradually damage blood vessels in the kidneys. Diabetes also damages nerves (see Chapter 14), which can result in difficulty emptying the bladder and which can cause pressure on the kidneys. When urine, especially urine with a high sugar content, remains in the bladder, infections become more common.

As the kidneys gradually lose their ability to function, levels of blood urea nitrogen (BUN) rise, along with levels of creatinine in the blood. Measurement of these two markers provides an accurate gauge of kidney function. In the following example of stage IV diabetically induced kidney disease, use of a Personal Tuner brought creatinine levels closer to normal and dramatically reduced insulin requirements. Don Goff explains:

> *I have been an insulin dependent diabetic for almost 10 years. In August of 2006 I was told I also had stage IV kidney disease and that I should be prepared in case my kidneys failed. I attended kidney dialysis classes and was entered into the kidney transplant program. I also began a special nutritional program. For the next two years my kidney function did not get any worse, but it did not improve with dietary intervention. Then early in 2009 I was introduced to the CAMS technology. I was given an opportunity to participate in a program where I was loaned a CAMS Personal Tuner. For three months I treated my kidneys twice a day for two minutes—but I saw no improvement. Then it was suggested that since my condition had progressed so far, I should increase the treatment time to ten minutes twice a day. When I did this, I began to see improvement. Over the next 3 months, my energy level improved and under my doctor's supervision, I was able to reduce my insulin requirement—by half. Subsequent tests indicated that kidney function was also improving. Creatinine levels dropped from 3.4 to 2.5 [normal levels are 0.6 to 1.2]. According to doctors,*

my kidney function has now stabilized at this new level—even without CAMS treatments which I have not had for six weeks.

For CAMS treatment of diabetically induced kidney disease, both the pancreas and the kidneys should be treated, along with any other obvious interference fields. For kidney disease resulting from other forms of damage to the blood vessels of the kidney, treatment should be directed to the kidneys and other offending interference fields.

Thyroid dysfunction

The thyroid, sometimes called the "master gland," has both metabolic and hormonal functions. It regulates metabolism by producing the thyroid hormones T3 and T4, which in turn regulate energy levels, body temperature, and weight. Few people have a clear idea of the vital role played by the thyroid in maintaining other functions, such as immune strength, emotional balance, the texture and quality of the skin and hair, and libido. According to Dietrich Klinghardt, MD, PhD, a neural therapist and developer of Autonomic Response Testing, thyroid disease is one of many conditions that respond well to treatment of interference fields.[1(p388)]

Thyroid dysfunction is now one of the most common endocrine weaknesses. It has been suggested that between 70% and 90% of North Americans suffer from minor to severe thyroid imbalance,[4] which contributes to the vast numbers of people who are overweight or suffer from eating disorders. Sluggishness, fatigue, and weight gain are all good indicators of suppressed thyroid function, otherwise known as hypothyroidism.

Hypothyroidism

When the thyroid gland is underactive, improperly formed at birth, or surgically removed, the affected person is said to have hypothyroidism. The most common symptoms are weight gain and fatigue, but other symptoms may also occur, including depression, sensitivity to cold, dry skin, muscle pain and weakness, and hoarseness. More fundamental factors that are seldom investigated in cases of hypothyroidism are toxicity (especially mercury toxicity), adrenal fatigue, food allergies, and celiac disease. Some of these factors, particularly the latter three, are suggestive of interference fields. When the adrenal glands are treated (as described on page 136) or when the digestive disturbance is addressed (as described in Chapter 12), some progress toward normalization of thyroid function may be made. Additionally, treatment of the thyroid itself may support the return of normal cell membrane electrical potential, as inflammation is brought under control. Inflammation is a major factor in Hashimoto disease, an underlying cause of hypothyroidism discussed in the next section.

Hashimoto disease

One of the most common factors in the development of hypothyroidism is a condition called Hashimoto disease. In this autoimmune condition, antibodies gradually target the thyroid and destroy its ability to produce thyroid hormones. Some people with Hashimoto disease develop nodules—benign and cancerous—in the thyroid gland. Because it is an inflammatory condition, Hashimoto disease can be alleviated with regular CAMS treatments. CAMS treatment may also be supportive in reducing nodules, as in this case shared by patient Barbara Reilly:

> *I have had Hashimoto's disease for over ten years. In my case, I developed numerous nodules on my thyroid which my doctor was watching to make sure they did not become cancerous. Following successful treatment using a TensCam to help with the passing of a kidney stone, I decided to treat my thyroid. The following month at a regular ultrasound scan, the doctor found no nodules on the left side of my thyroid. They ran the scan three times to make sure. With that good news, I decided to continue to treat my thyroid (2–3 times a month). At my next*

intment, tests revealed that my thyroid levels were ve also experienced more energy and a greater -being since I began to use the TensCam for my thyroid.

Hyperthyroidism

When the thyroid gland becomes overactive and produces too much thyroid hormone, the affected person is said to have hyperthyroidism. Higher-than-normal levels of thyroid hormones increase the metabolism, causing weight loss. The most common cause of hyperthyroidism is the autoimmune condition known as Graves disease.

Graves disease

Graves disease occurs when the immune system mistakenly attacks the thyroid gland and causes it to overproduce thyroid hormones. The overstimulated thyroid often becomes enlarged, producing a goiter. Treatments for Graves disease and other forms of hyperthyroidism are directed at slowing the production of thyroid hormones. Radioactive iodine, thyroid-suppressing drugs, and surgery are options, along with immune suppressants and/or anti-inflammatory drugs. CAMS treatments offer yet another choice. Controlling the inflammatory component of Graves disease and restoring normal membrane electrical potential can reduce symptoms—often to a greater extent than drugs or surgery— without adverse effects. Treatment of the offending interference field can sometimes resolve symptoms entirely. In such cases, the thyroid and any other offending interference fields should be treated directly.

Adrenal dysfunction

The adrenal glands, located on top of each kidney, produce hormones that participate in many bodily functions. One of these hormones is cortisol, which affects every organ and every tissue. The adrenal glands also produce aldosterone and androgens. Aldosterone regulates electrolytes, thereby affecting blood volume and blood pressure. Androgens are sex hormones, the most important of which is dehydroepiandrosterone (DHEA).

Adrenal fatigue

One of cortisol's most important functions is to help the body respond to stress. If stress remains high for prolonged periods, the adrenals are forced to produce cortisol continuously. Over time, they become unable to keep up with the demand. This condition is referred to as adrenal fatigue or adrenal overload. A high percentage of individuals in the United States experience undiagnosed adrenal fatigue, which has a domino effect on other hormones. When the adrenals are chronically overworked, they produce less DHEA. This affects the hormonal balance of the entire body. Insufficient DHEA contributes to fatigue, loss of strength and muscle mass, depression, aching joints, decreased sex drive, and impaired immune function. For many who are under continual stress—who may exhibit any of the above symptoms—treating the adrenal glands with a CAMS device can make a big difference. CAMS treatment re-establishes order in the bioenergy field, calming tissues, normalizing cell membrane electrical potential, and reducing inflammation. As has been mentioned many times in this book, these factors are the cause—at several different levels—of disease and discomfort.

Addison disease

Addison disease, sometimes called chronic adrenal insufficiency, occurs when the adrenal glands cannot produce enough of hormones. Autoimmune destruction of the adrenals is the most common form of Addison disease in the United States, accounting for 80% of all cases. Addison disease has many of the same symptoms as adrenal fatigue. The outcome is the same: too few adrenal hormones. Treatment of Addison disease (by directing the CAMS device at the adrenal glands or other related interference fields) can often reverse the symptoms and reinstate youthful stamina and strength.

References

1. Trivieri L, Anderson J, editors. *Alternative medicine: the definitive guide*. 2nd ed. Celestial Arts Publishing Co; 2002.
2. Dosch JP, Dosch M. *Manual of neural therapy according to Huneke*. 2nd English ed. (translation of 14th German ed.). Gutberlet R, translator. Thieme Medical Publishers; 2007.
3. Diabetes and kidney disease. National Kidney Foundation. Available from: *http://www.kidney.org/atoz/content/diabetes.cfm*
4. Langer S, Scheer J. *Solved: the riddle of illness*. Keats Publishing; 2000.

Chapter 16

Psychological and Emotional Conflicts

The connection between physical problems and their psychological or emotional factors has been mentioned repeatedly in the preceding chapters. Although this relationship has been recognized intuitively for centuries, science has finally given us the evidence to connect the dots between them. The many books that have been written on the subject have been pivotal in helping us to understand that the energy of traumatic emotional events often gets "stuck" in the bioenergy field. When emotional events remain unresolved, the disturbance in the energy field eventually causes a related disturbance in the tissues of the body. Sooner or later, physical problems develop.

It is especially difficult to make the connection between emotional events and physical problems when long periods have passed between the precipitating event and the development of symptoms. The amount of time is usually longer for events experienced as a child than for events experienced as an adult. Because a child's body is more resilient than that of an adult, it may take 20 years or more for the physical symptoms of a childhood trauma to become evident. However, behavioral problems often manifest earlier. Such symptoms, either physical or behavioral, are indications of an interference field. They are a message from the body that something is wrong, in the same way that pain is a message indicating the existence of a problem.

Many traumatic events occur when we are children. Some can be extremely traumatic, such as abuse. Others may seem silly when viewed from the adult perspective, but for the child, who lacks adult experience and adult wisdom, even apparently minor events can seem overwhelming. Children often blame themselves for traumatic events in which they have no role, such as quarrels between parents or accidents. Each of these leaves an emotional scar, not dissimilar to the physical scars that can cause interference fields.

Adults experience emotional traumas too, such as the loss of a loved one or the betrayal of a friend. For an adult, the time lag between emotional trauma and physical symptoms is not nearly so long. Symptoms may begin to occur within days or weeks of the traumatic event.

Unresolved emotional conflicts usually involve a contradiction—an unsolved problem that leads to or that supports an existing faulty belief pattern. When the contradiction cannot be resolved, the subconscious mind moves it to the background (into the subconscious mind), so that we can continue to function in our everyday lives. This makes unresolved emotional conflicts even harder to identify and resolve, especially without an understanding of the ways in which interference fields hold perturbed energy patterns.

The medical profession currently has no way of identifying energetic disturbances (scars) in the bioenergy field, nor does it have any way of treating them. This may explain the hit-or-miss approach to using drugs and surgery. No amount of drugs or surgery can overcome a disturbance in the bioenergy field, and no treatment of one area of the body can compensate for a disturbance that has its origin elsewhere. Drugs or surgery may mask the problem, but the energetic disturbance will eventually cause symptoms in another way or in another part of the body.

Associations between organs and emotions

Traditional Chinese Medicine associates certain emotions with specific organs. Astute Western practitioners regularly observe these relationships. When an emotionally charged experience is not resolved appropriately, the emotion sometimes expresses itself in a physical way through the corresponding organ. A list of organs and their corresponding emotions follows:

Bladder: shame, self-pity, hurt

Gallbladder: resentment, blame, bitterness

Heart: betrayal, trust, disappointment, lack of joy, self-protection

Kidney: fear, guilt, powerlessness

Large intestine: control, criticism

Liver: anger, frustration

Lungs: grief, sorrow, despair

Pancreas: inadequacy, self-worth

Small intestine: loneliness, abandonment

Stomach: anxiety, stress, hate

Thyroid: repression, self-expression

These organ–emotion associations can sometimes be helpful in deciding where to direct CAMS treatment. When emotional conflicts are suspected, it is never inappropriate to treat the associated organ. As has been discussed previously, there is a spot located on the top of the head, above and behind the right ear, that often manifests as an interference field when the person has unresolved emotional conflict. Treatment of this area should always be considered. In my own clinical experience, 75% to 80% of illnesses can be helped through treatment of this single area. When the practitioner suspects an unresolved emotional issue, treating the spot on the top of the head along with the organ that is most closely associated with the emotion is a good place to start. Often the organ itself will be the point of physical disturbance.

The following example shows how emotional issues create energetic disturbances and physical symptoms. It also illustrates how treatment of an offending interference field can result in the profound resolution of pain and other physical symptoms. This interesting case occurred during a demonstration of the CAMS technology at a physician's office. The physician had invited patients whose treatment was proving particularly difficult to see if the CAMS technology could help where other treatments had failed.

> *A 54-year-old woman had begun to experience symptoms of low back and sciatic pain at the time of her father's death 20 years earlier. Eventually, she also developed foot drop on her right side. This woman had undergone three different surgical procedures in an attempt to relieve the pain. Although unsuccessful, the surgeries had eliminated the possibility of a ruptured disc and of pinched nerves. No cause for the pain could be identified. Manual energy evaluation revealed an interference field on the top left portion of her head (the homunculus). A two-minute TensCam treatment completely resolved all pain. The woman also walked out of the office normally—with no evidence of foot drop.*

Conclusion

In classical medicine today, disease is defined by its final manifestation. In many cases, diagnostic methods fail to recognize any of the precursors of the diseased state until the final expression of its presence. Drugs and other treatments are then directed at the end result. This method of diagnosis and treatment simply identifies and then suppresses the symptoms. It may give the appearance of a cure, but it does not address the underlying cause, which continues to go unrecognized and untreated, usually with further consequences down the road.

As sophisticated as our modern medical system has become, it still offers only band-aid solutions. For true healing to take place, the cause—not the end result—must be identified and addressed. This has been the foundation of osteopathic medicine from its origins. Not only are the students of osteopathic medicine taught the standard medical curriculum, they are also taught the interrelationships among the body's nerves, muscles, bones, and organs. They are additionally trained in osteopathic manipulative treatment to assist the body's healing process. Osteopathic doctors and many other healthcare professionals are now trained to treat the whole person and to look for the underlying cause of illness. This is a step in the right direction.

However, in attempting to identify the real cause, it is sometimes difficult to keep from being sidetracked. Some conditions *appear* to have a direct cause (a viral, bacterial, or parasitic infection), but the true cause lies at a deeper level. Practitioners performing traditional diagnosis get caught in the trap of identifying "a causative agent" while paying little attention to the status of the internal environment that allowed the causative agent to take hold in the first place. It is well understood that of many individuals who are exposed to the same virus or bacterium, only some become infected. The question "Why?" opens the door to serious investigation of what has been referred to as the "biological terrain"—the condition of the internal environment.

Claude Bernard, a contemporary of Louis Pasteur, developed the theory that the body's ability to resist disease was dependent on its general condition or on its biological terrain—the interior milieu. Because of political conflicts, Bernard's work was disregarded by most physicians, who instead favored antibiotics and vaccines, thus fueling the ultimate growth of the pharmaceutical industry. Only during the past several decades has research uncovered the basis for Bernard's conclusions. This research has brought forth a new understanding of the connection between disease and the biological terrain, a process that likens the terrain to good soil. When the soil is healthy and balanced, nutritive crops (health) can be successfully cultivated. When the soil is deficient, only weeds (disease) will grow.

The recent focus on the biological terrain, by some in the medical profession and by many alternative healthcare practitioners, brings us closer to the mark, yet it is still one step removed from the true cause. Only when we grasp the significance of the bioenergy field will we truly understand the nature of health and healing. And only when we comprehend our connection with the Earth and with the subtle energies available in nature will we be able to create ongoing, vibrant health.

In the previous pages, I have presented evidence that the cause of disease resides in the bioenergy field. Following Bernard's line of thinking, we begin to see the bigger picture. Understanding that the biological terrain extends beyond the physical body takes us to the ultimate cause of disease. It is the condition of the *energetic* terrain that determines the condition of the *physical* terrain, and this in turn determines the health of the cells. In other words, disturbances in the human energy field are the ultimate cause of pain, anxiety, fatigue, and other manifestations of dis-ease. Conversely, maintenance of the integrity (coherence) of the bioenergy field results in health.

It is the condition of the *energetic* terrain that determines the condition of the *physical* terrain, and this in turn determines the health of the cells.

In the foregoing text, I have also identified the importance of the Earth's supportive energy field—the "tuning fork" that harmonizes all life on the planet. This recognition leads to an understanding of certain qualities of the Earth's most prevalent substance, quartz, and the significance of its crystalline form for amplifying and converting electromagnetic energy to scalar energy. This scalar energy is considered by some to be the energetic language of the universe.

I hope that this book has also revealed the power of intention as an organizing force, as well as the sea of energy that exists everywhere—the source of and the avenue through which all information in the universe is conducted. These pieces are the necessary keys to an advanced method of treatment that can effectively change the energetic terrain, re-establishing coherence in the bioenergy field, with subsequent healthful changes at all levels in the physical body.

Conventional medicine versus energy medicine

A growing number of individuals have decided that the risks of conventional medicine are too high, that there are too many adverse effects and too many potential complications. For this and other reasons, there has been a tremendous surge in the demand for complementary and alternative healing. Therapies like acupuncture, homeopathy, Reiki, and Ayurveda, among many others, are growing in popularity. More than half of the US population now relies on complementary therapies to support traditional care. Often these nontraditional methods are tried first, before the patient seeks conventional therapy. Unfortunately, in many other instances, they are explored as a last resort.

The ancient traditions of qigong and acupuncture, among others, have always assumed the existence of energy that Western scientific instruments is only now beginning to detect. Many of these ancient methods are noninvasive, dealing only with the body's energy. Their track record has survived hundreds, if not thousands, of years. Most of these complementary therapies use a model that includes an understanding of the energetic component of the human body—the connection between the body, the mind, and the spirit via the brain and the nervous system. In contrast, Western culture has only recently begun to accept the idea that our bodies are made of energy. Even conventional medicine is just beginning to comprehend the emotional and the mental components of illness.

Without a doubt, human beings are more than a physical casing and the glands and organs enveloped by that casing. Learning to work with energy and to focus treatment on the energetic body is the leap that Robert Fulford predicted for 21st century medicine. Certainly he understood the advantages of noninvasive energy medicine as it relates to the Hippocratic Oath and the dictum to "first, do no harm."

Success with CAMS technology

CAMS technology is a breakthrough in medical care. Not a single adverse effect has been reported in over 10 years, since the CAMS technology first became available. Even for the first-time user, CAMS treatment can do no harm. In an era when drugs have a success rate of about 40% (depending on who is defining the word "success") and surgery has a success rate no better than 60%, the success rate of 80% or more achieved by experienced users of the CAMS technology is highly encouraging. Heathcare practitioners are using it not only for their patients but also for themselves. William Halcomb, DO, of Mesa, Arizona, has this to say about the device:

> *In over 50 years of practice, the TensCam is the most valuable piece of equipment that I have ever had. Also, for personal health problems, both my wife and I have used it extensively with positive results.*

When you consider that an alarming one of every four patients suffered observable adverse effects from the more than 3.3 billion prescriptions filled in the year 2002[1] and that 2.2 million adverse drug reactions were reported in US hospitals in 2003,[1] the CAMS treatment option is looking better all the time.

Reference

1. Null G, Dean C, Feldman M, Rasio D, Smith D. Death by medicine. *Life-Extension Mag* 2006 Aug. Available from: *http://www.lef.org/magazine/mag2006/aug2006_report_death_08.htm*

Appendix

Clinical Research at the University of Central Florida

In a double-blind, placebo-controlled clinical study conducted in 2002 at the University of Central Florida, the TensCam device was tested for the relief of low-back pain. This work revealed the safety and effectiveness of the CAMS technology. Ninety-eight patients with low-back pain resulting from a sprain, strain, or other injury participated in the study.

Methods

The participants were randomly divided into two groups. The test group received CAMS treatments delivered to the area of pain from a distance of 18–24 inches; the control group received sham treatments from an identical-looking placebo device. Neither the participants nor the researchers administering the treatments knew which CAMS device was real and which was the placebo.

Before treatment, the participants were asked to rate their level of pain on the 10-point visual analog scale. The participants were asked to re-evaluate their level of pain 10 minutes after the treatment and again 24 hours later.

The participants were asked not to take any over-the-counter or prescription pain relievers beginning 12 hours before the treatment until after the 24-hour post-treatment evaluation.

Results

Twenty-nine (60%) of the test group of 48 participants who received the actual CAMS treatment reported improvement 10 minutes after treatment, and 30 (63%) reported improvement at the 24-hour evaluation.

Eighteen (36%) of the control group of 50 participants who were treated with the placebo device showed improvement 10 minutes after treatment, but the rate of improvement declined to 22% (11 participants) at the 24-hour evaluation.

No complications or adverse effects were reported. The full report is available online at *http://www.tenscam.com/content/tenscam_clinical_research.pdf*

About the Author

Charles J. Crosby, DO, MD(H), is a retired orthopedic surgeon with more than 30 years of clinical experience. He graduated in 1969 from the Kirksville College of Osteopathic Medicine and eventually became a fellow of the American Academy of Osteopathy and the American Academy of Osteopathic Specialists. He is board-certified in three specialty areas—in orthopedic surgery, in osteopathic manipulative medicine, and in pain management. Dr. Crosby is additionally trained in homeopathy and is recognized by the state of Arizona as a homeopathic medical doctor.

Dr. Crosby was in private practice for 15 years as an orthopedic surgeon before accepting a position in Orlando, Florida, as a pain management specialist for the Veterans Administration (VA). While at the VA, he had the opportunity to do a significant amount of research and to explore many avenues for the management of pain. He and two other doctors established an 86% success rate and were able to reduce the use of pain medications by outpatients at the clinic by 45%. Much of their success was due to osteopathic manipulative treatment (the correction of skeletal misalignments) and to the incorporation of other modalities that are often referred to as "energy medicine." This unprecedented track record convinced Dr. Crosby that he was on the right track. When he returned to private practice, he was determined to make a difference on a much larger scale. His attendance at Dr. Fulford's last lecture in 1997 set the stage for him to make the contribution he had envisioned for many years.

The CAMS technology took several years to develop—and even longer to fully understand. As with most technological advances, it is a work still in progress. Crosby has conducted dozens of workshops around the world for clinicians and patients in an effort to introduce and to advance the CAMS technology. He is a member of the following organizations:

- American Osteopathic Association
- Florida Osteopathic Medical Association
- American Academy of Orthopaedic Medicine
- Cranial Academy